Place-Value Chart (Whole Numbers)

2	3	6	,	5	1	8	,	9	0	4
hundred millions	ten millions	millions		hundred thousands	ten thousands	thousands		hundreds	tens	ones

AF855842

Place-Value Chart

4	.	8	1	6	2	3	9	4	5	7	0	6
		tenths	hundredths	thousandths	ten-thousandths	hundred-thousandths	millionths	ten-millionths	hundred-millionths	billionths	ten-billionths	hundred-billionths

Points on a Number Line

The tick marks on each number line are equally spaced. Which letter (if any) corresponds to the given number? Which number corresponds to the given letter?

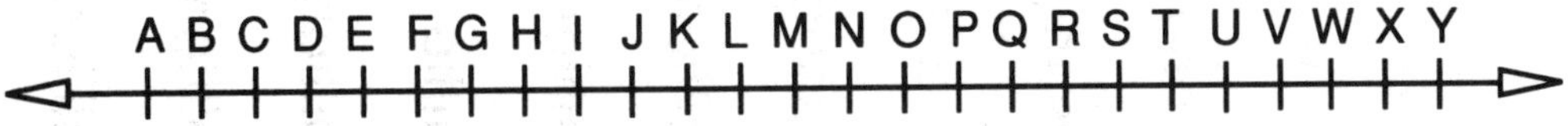

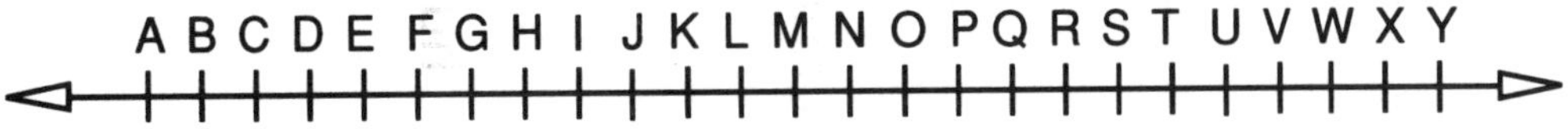

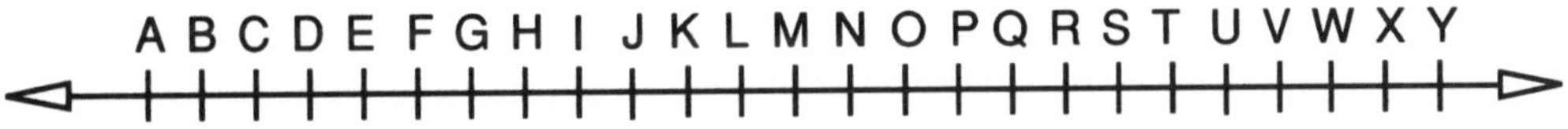

Number Lines

How would you graph the given numbers on the number line?

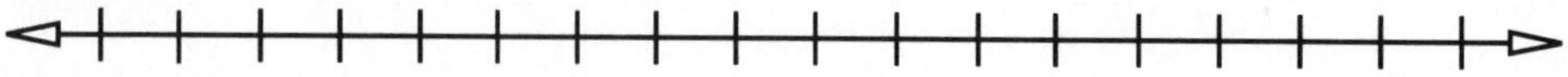

Question 29

The thermometer pictured below shows Joanne's body temperature on three consecutive days of a cold. Put the three numbers into one sentence connected by inequality symbols.

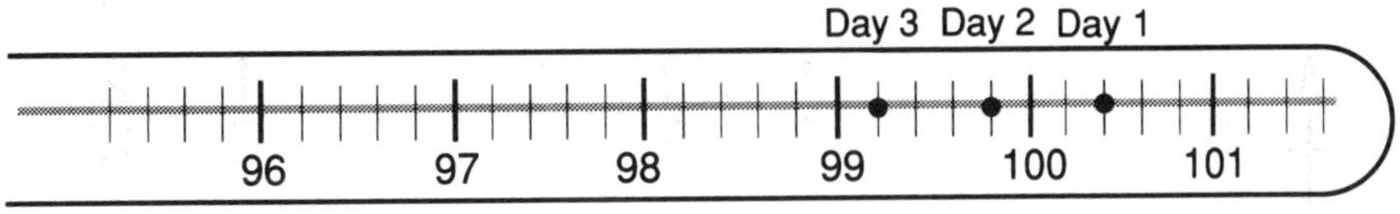

Additional Question

On the unmarked thermometer, graph the given three temperatures. Then write the three numbers in one sentence connected by inequality symbols.

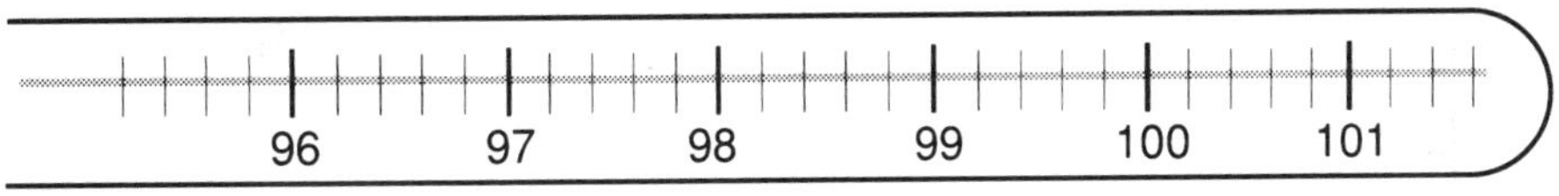

Common Fraction-Decimal Equivalents

Fourths and Eighths

$\frac{1}{8} =$

$\frac{1}{4} = \frac{2}{8} =$

$\frac{3}{8} =$

$\frac{2}{4} = \frac{4}{8} =$

$\frac{5}{8} =$

$\frac{3}{4} = \frac{6}{8} =$

$\frac{7}{8} =$

Thirds and Sixths

$\frac{1}{6} =$

$\frac{1}{3} = \frac{2}{6} =$

$\frac{3}{6} =$

$\frac{2}{3} = \frac{4}{6} =$

$\frac{5}{6} =$

Fifths and Tenths

$\frac{1}{10} =$

$\frac{1}{5} = \frac{2}{10} =$

$\frac{3}{10} =$

$\frac{2}{5} = \frac{4}{10} =$

$\frac{5}{10} =$

$\frac{3}{5} = \frac{6}{10} =$

$\frac{7}{10} =$

$\frac{4}{5} = \frac{8}{10} =$

$\frac{9}{10} =$

Fractions and Decimals on a Unit Segment

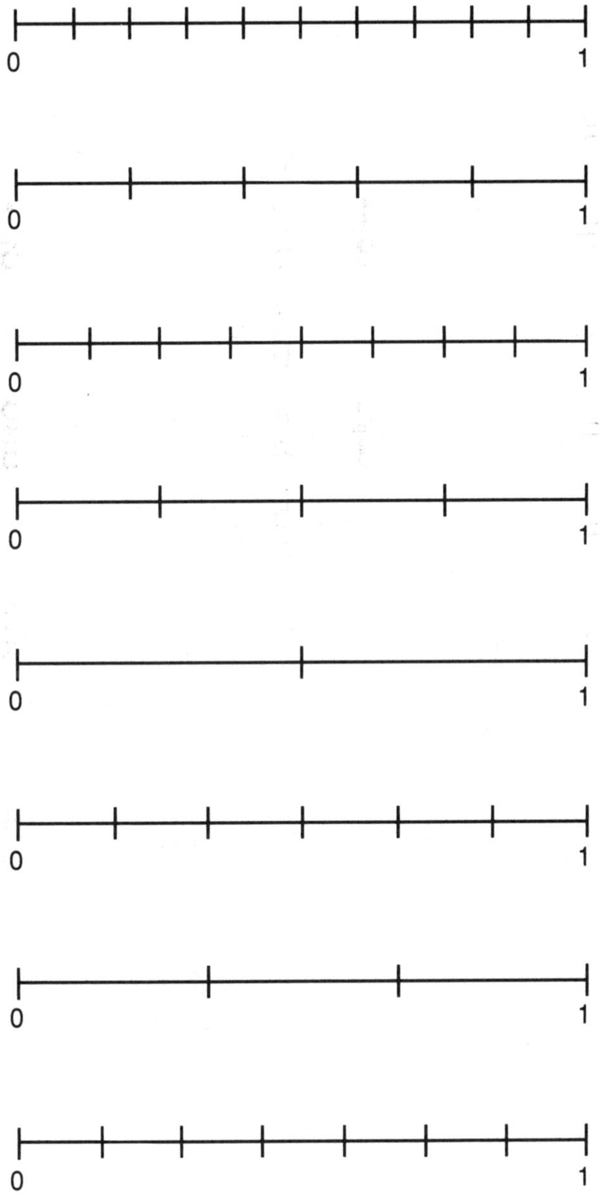

World Population

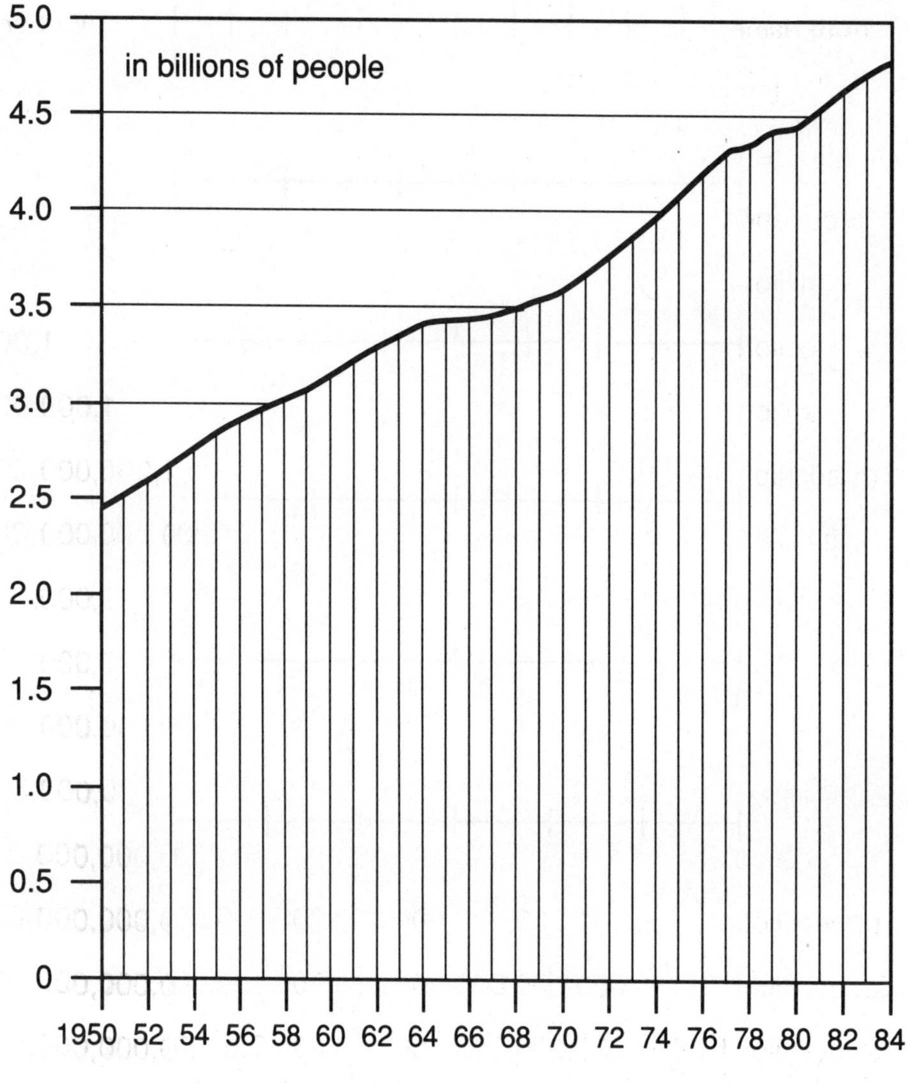

Source: Center for International Research, U. S. Bureau of Census

Powers of 10

Power of 10	Word Name	Written as decimal
10^{1}	ten	10
10^{2}	hundred	100
10^{3}	thousand	1,000
10^{6}	million	1,000,000
10^{9}	billion	1,000,000,000
10^{12}	trillion	1,000,000,000,000
10^{15}	quadrillion	1,000,000,000,000,000
10^{18}	quintillion	1,000,000,000,000,000,000
10^{21}	sextillion	1,000,000,000,000,000,000,000
10^{24}	septillion	1,000,000,000,000,000,000,000,000
10^{27}	octillion	1,000,000,000,000,000,000,000,000,000
10^{30}	nonillion	1,000,000,000,000,000,000,000,000,000,000
10^{33}	decillion	1,000,000,000,000,000,000,000,000,000,000,000
10^{36}	undecillion	1,000,000,000,000,000,000,000,000,000,000,000,000
10^{39}	duodecillion	1,000,000,000,000,000,000,000,000,000,000,000,000,000
10^{42}	tredecillion	1,000,000,000,000,000,000,000,000,000,000,000,000,000,000

Powers of Two

2^1	2	2^{14}	16,384
2^2	4	2^{15}	32,768
2^3	8	2^{16}	65,536
2^4	16	2^{17}	131,072
2^5	32	2^{18}	262,144
2^6	64	2^{19}	524,288
2^7	128	2^{20}	1,048,576
2^8	256	2^{21}	2,097,152
2^9	512	2^{22}	4,194,304
2^{10}	1024	2^{23}	8,388,608
2^{11}	2048	2^{24}	16,777,216
2^{12}	4096	2^{25}	33,554,432
2^{13}	8192	2^{26}	67,108,864

Representing Percents on a Number Line

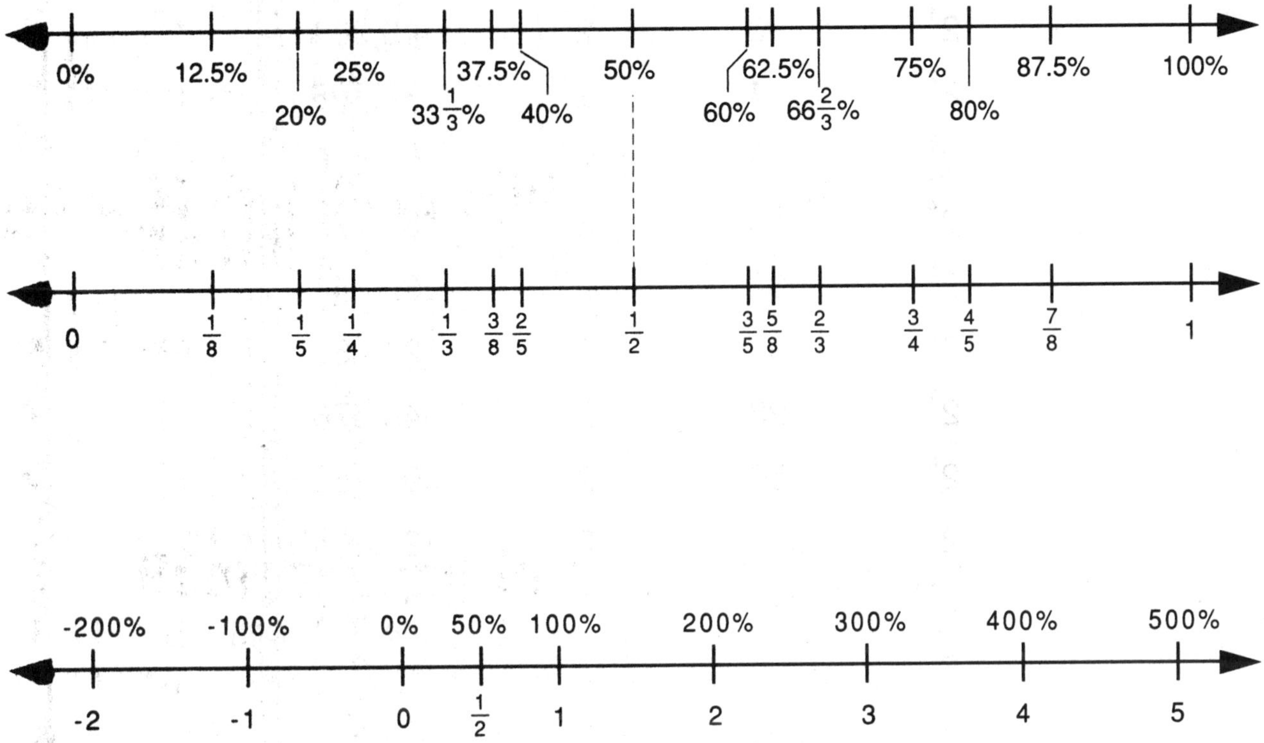

Positive and Negative Powers of 10

The names of the negative powers of 10 follow closely those for the positive powers.

10^{1}	= ten	10^{-1}	= one tenth
10^{2}	= one hundred	10^{-2}	= one hundredth
10^{3}	= one thousand	10^{-3}	= one thousandth
10^{6}	= one million	10^{-6}	= one millionth
10^{9}	= one billion	10^{-9}	= one billionth
10^{12}	= one trillion	10^{-12}	= one trillionth
10^{15}	= one quadrillion	10^{-15}	= one quadrillionth
10^{18}	= one quintillion	10^{-18}	= one quintillionth

Measuring Length

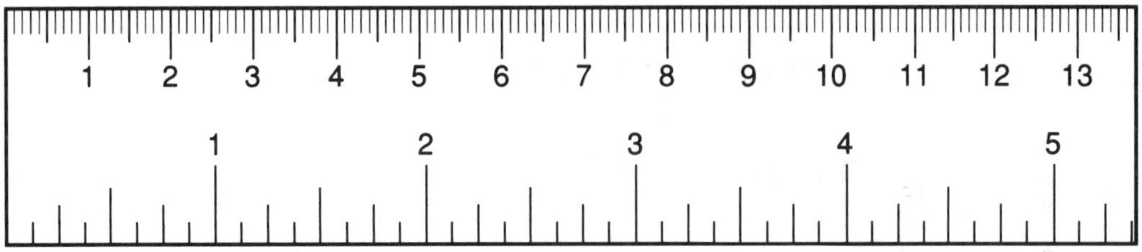

Measure each of the following segments using the specified units.

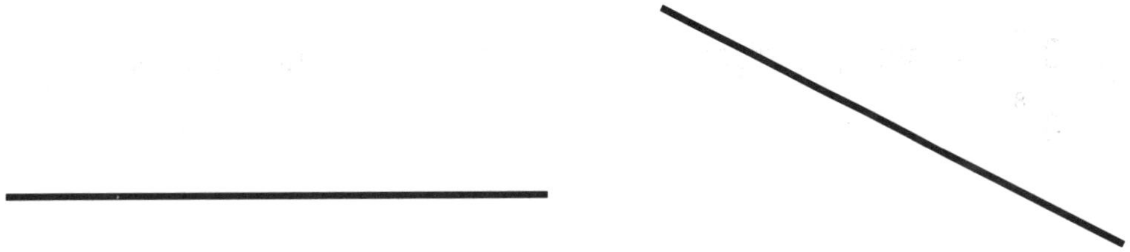

(a) nearest centimeter ______

(b) nearest tenth of a cm ______

(c) nearest inch ______

(a) nearest inch ______

(b) nearest $\frac{1}{2}$ inch ______

(c) nearest $\frac{1}{8}$ inch ______

Common Units in the U.S. System of Measurement

For length:	1 foot (ft) = 12 inches
	1 yard (yd) = 36 feet
	1 mile (mi) = 5280 feet
For weight:	1 pound (lb) = 16 ounces (oz)
	1 short ton = 2000 pounds
For capacity (liquid or dry volume):	1 quart (qt) = 2 pints
	1 gallon (gal) = 4 quarts

The International Metric System of Measurement

Table of Prefixes

place value	thousands	hundreds	tens	ones	tenths	hundredths	thousandths
power of 10	10^3			10^0			
unit of length	**kilometer**					**centimeter**	
unit of mass		hectogram			decigram		
unit of capacity			dekaliter				**milliliter**

place value	trillions	billions	millions
power of 10	10^{12}		10^6
prefix	tera-	giga-	mega-

millionths	billionths	trillionths
	10^{-9}	10^{-12}
micro-	nano-	pico-

Five Conversions You Need to Know

1 inch = 2.54 centimeters (exactly)

1 kilogram ≈ 2.2 pounds

1 liter ≈ 1.06 quarts

1 kilometer ≈ 1.6 miles

1 meter ≈ 39.37 inches

For which conversion(s) can you use the Substitution Principle?

For which conversion(s) should you use the Estimation Principle?

Measuring Angles

Measure each angle.

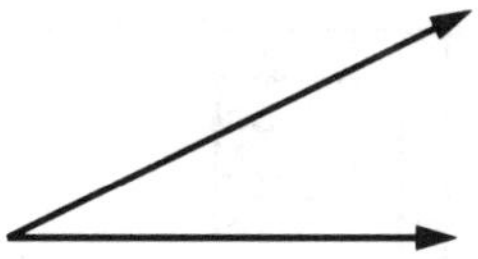

________°

________°

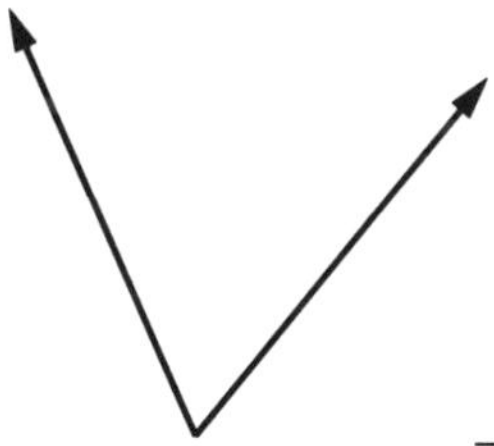

________°

________°

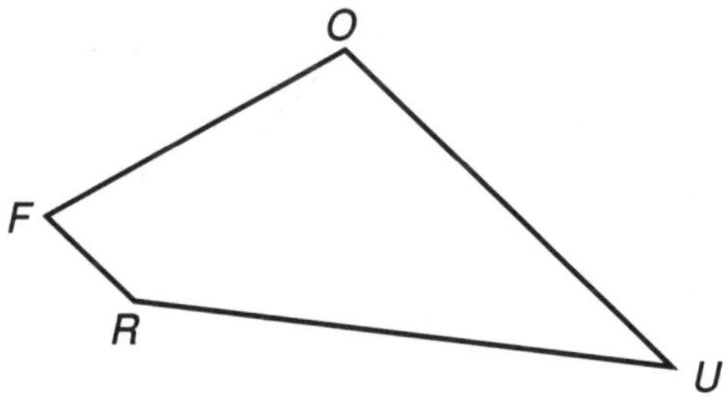

m∠ F = ________

Square Foot

1 square foot = ______ square inches.

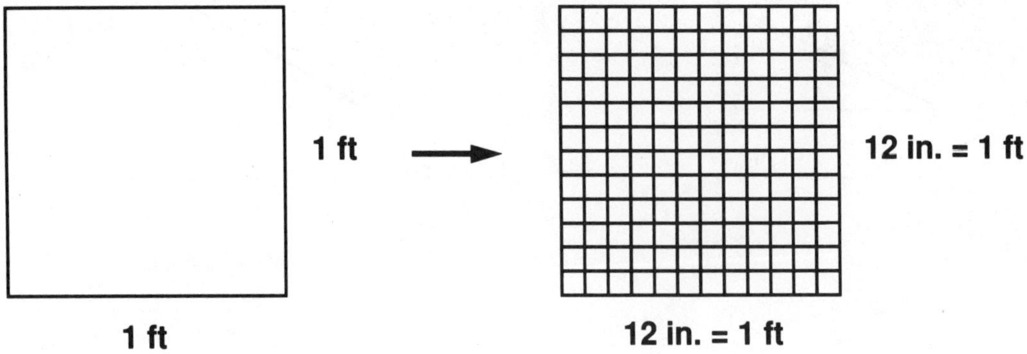

Cubic Foot

1 cubic foot = ______ cubic inches.

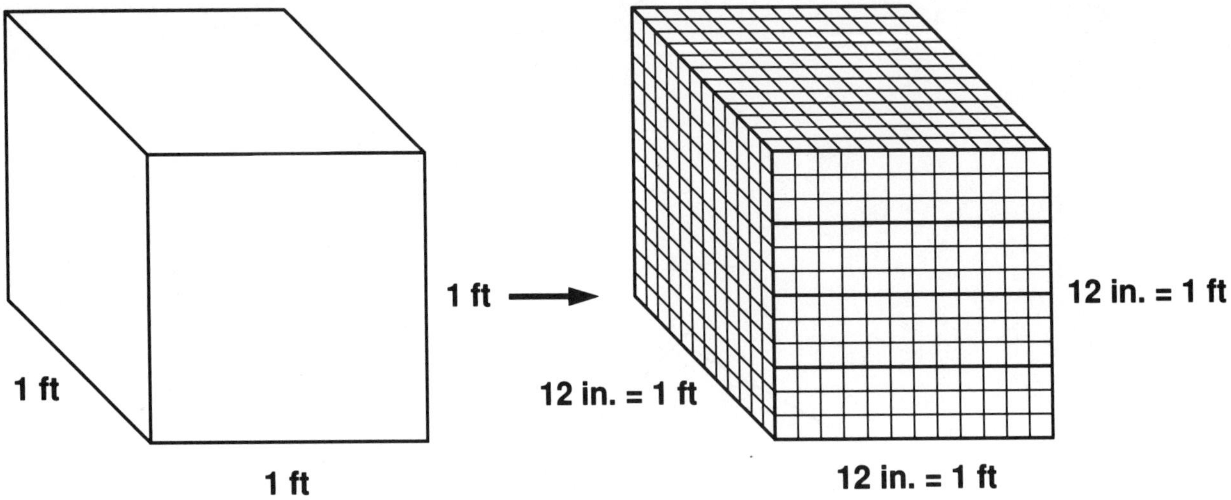

Making Cubes

Copy the patterns. Draw the side of each square to equal 1 inch. Fold along the dotted lines in each pattern, then tape to form a 1-inch cube. Use the patterns to make several cubes.

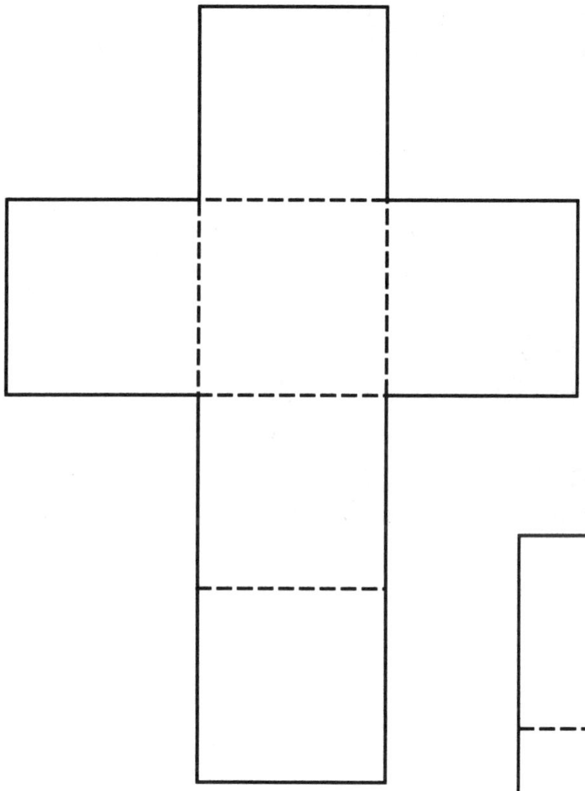

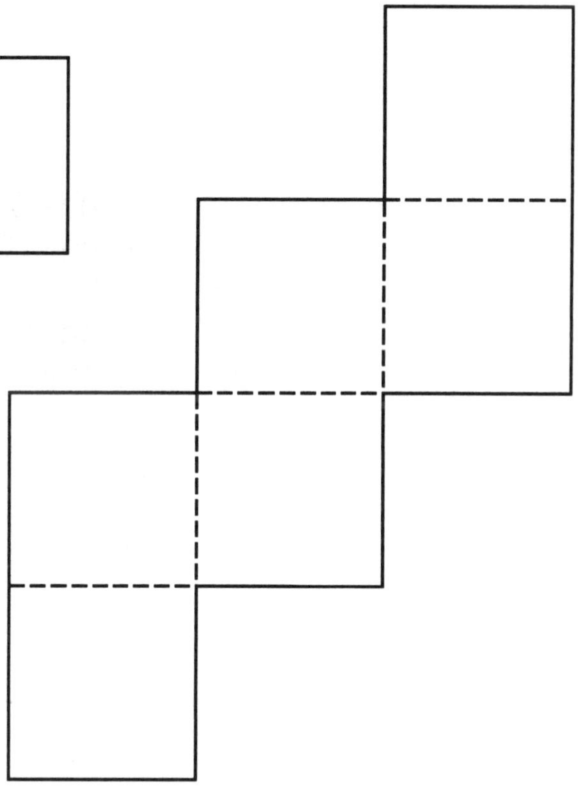

Describing Patterns with Variables

$5\% = 5 \times .01$
$43.2\% = 43.2 \times .01$
$78\% = 78 \times .01$
$n\% = n \cdot .01$

$1.43 + 2.9 = 2.9 + 1.43$
$12 + 37 = 37 + 12$
$\frac{8}{3} + \frac{7}{5} = \frac{7}{5} + \frac{8}{3}$
$a + b = b + a$

1 person has $2 \cdot 1$ eyes.
2 people have $2 \cdot 2$ eyes in all.
3 people have $2 \cdot 3$ eyes in all.
4 people have $2 \cdot 4$ eyes in all.
p people have $2 \cdot p$ eyes in all.

Additional Examples

1. Which rectangles have the same area?

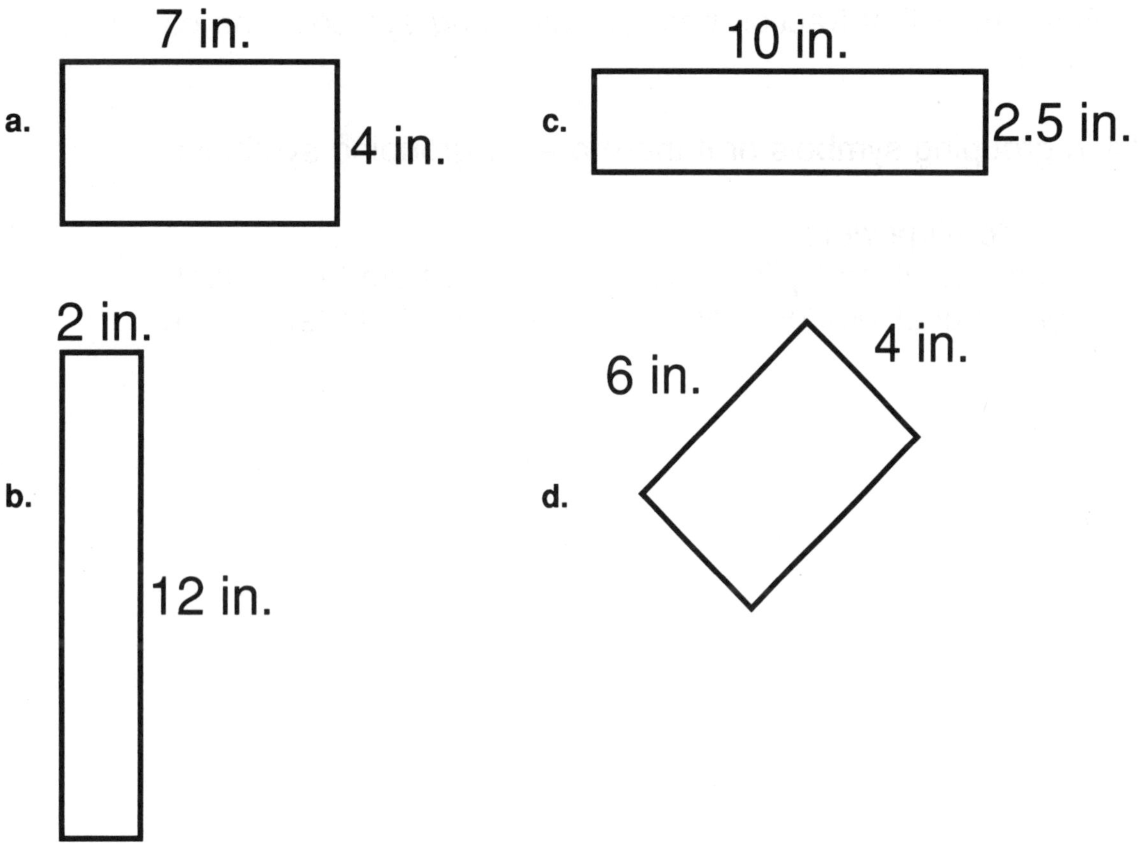

2. Find the area of the rectangle in square centimeters.

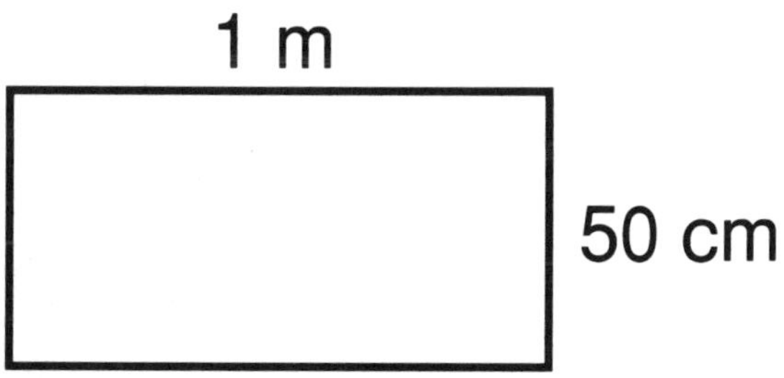

Summary of Rules for Order of Operations

1. First do operations within parentheses or other grouping symbols.

A. If there are nested grouping symbols, work in the innermost symbols first.
B. Remember that fraction bars are grouping symbols and can be different from /.

2. Within grouping symbols or if there are no grouping symbols:

A. First, do all powers.
B. Second do all multiplications or divisions, from left to right.
C. Then do additions or subtractions in order, from left to right.

Questions 18–21

Match each sentence with the graph of its solutions.

18. $1 \le m \le 4$
19. $1 < m < 4$
20. $1 < m \le 4$
21. $1 \le m < 4$

(a)

(b)

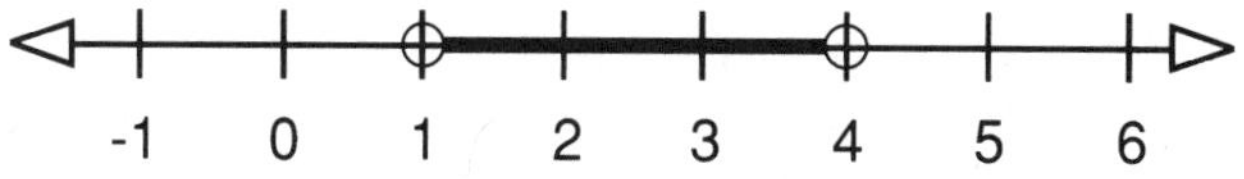

(c)

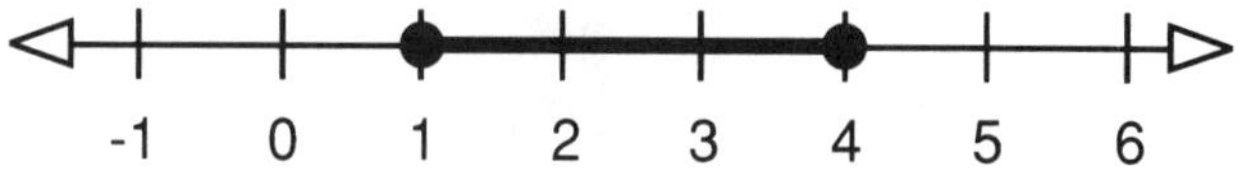

(d)

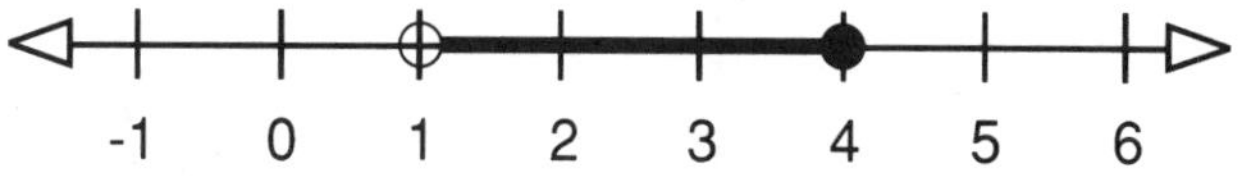

Question 34

		2	13
5		11	
	6		12
4	15	14	

Examples 2 and 3

2. -4 + 7 = ________

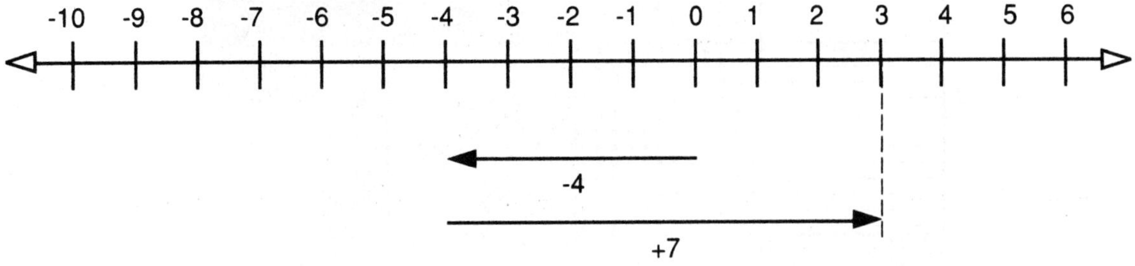

3. -3 + -2 = ________

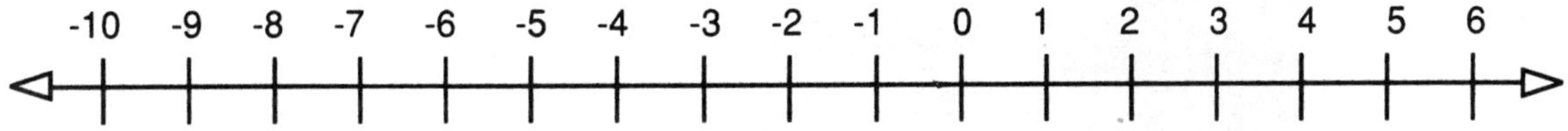

Questions 45 and 46

45.

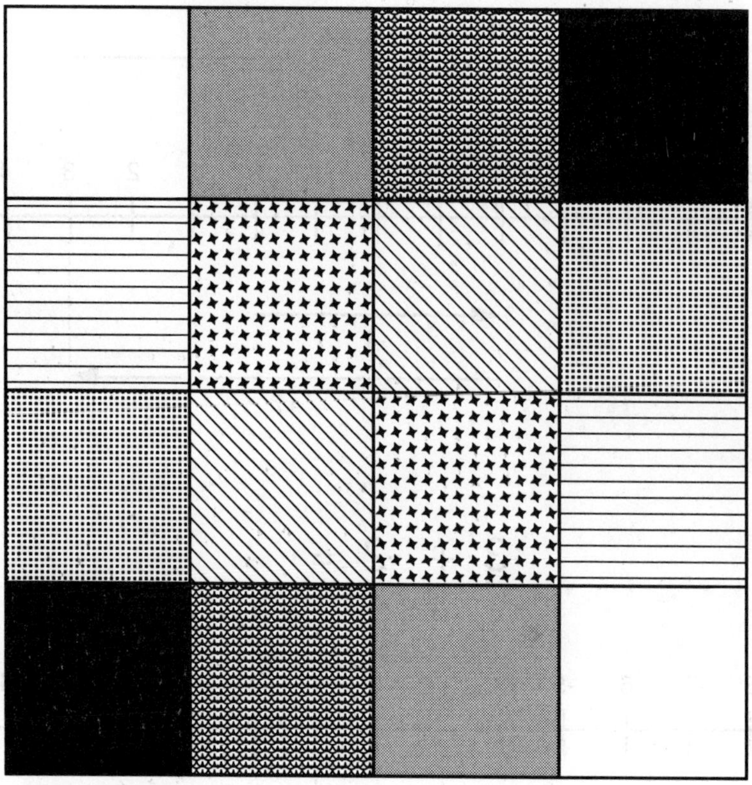

46.

Squares	Number
4 × 4	
3 × 3	
2 × 2	
1 × 1	
Total	

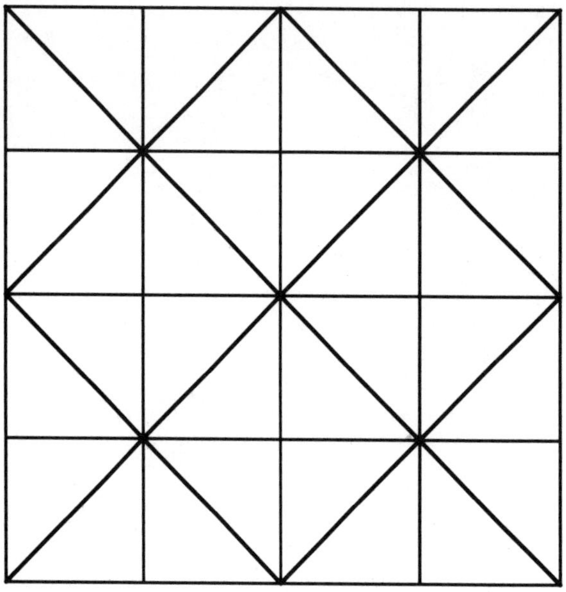

Rotations

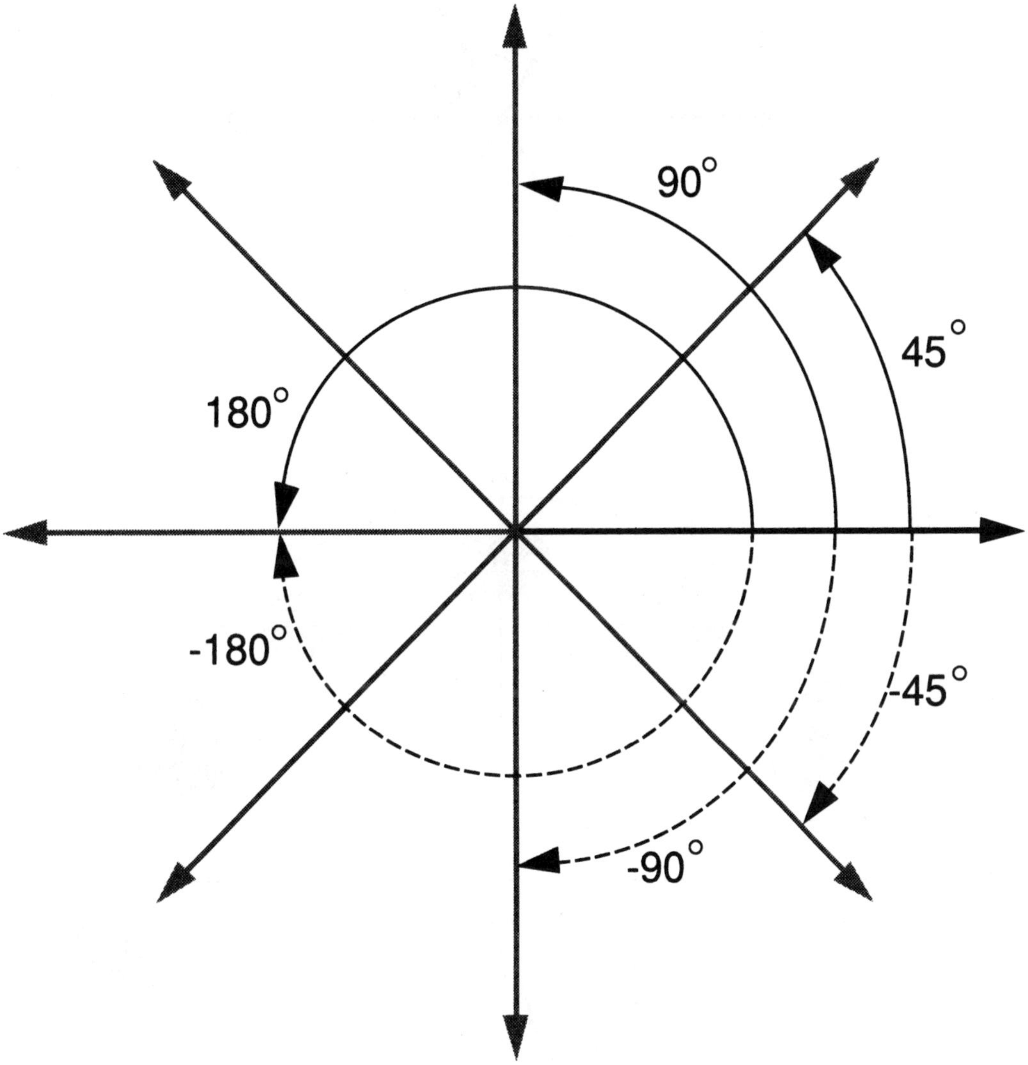

Questions

8–11.

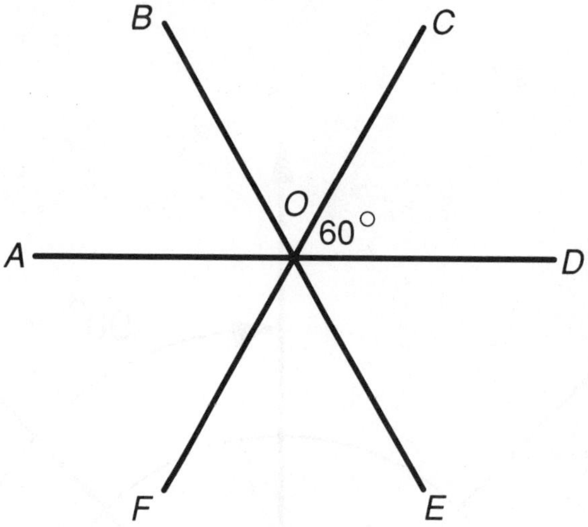

21–23.

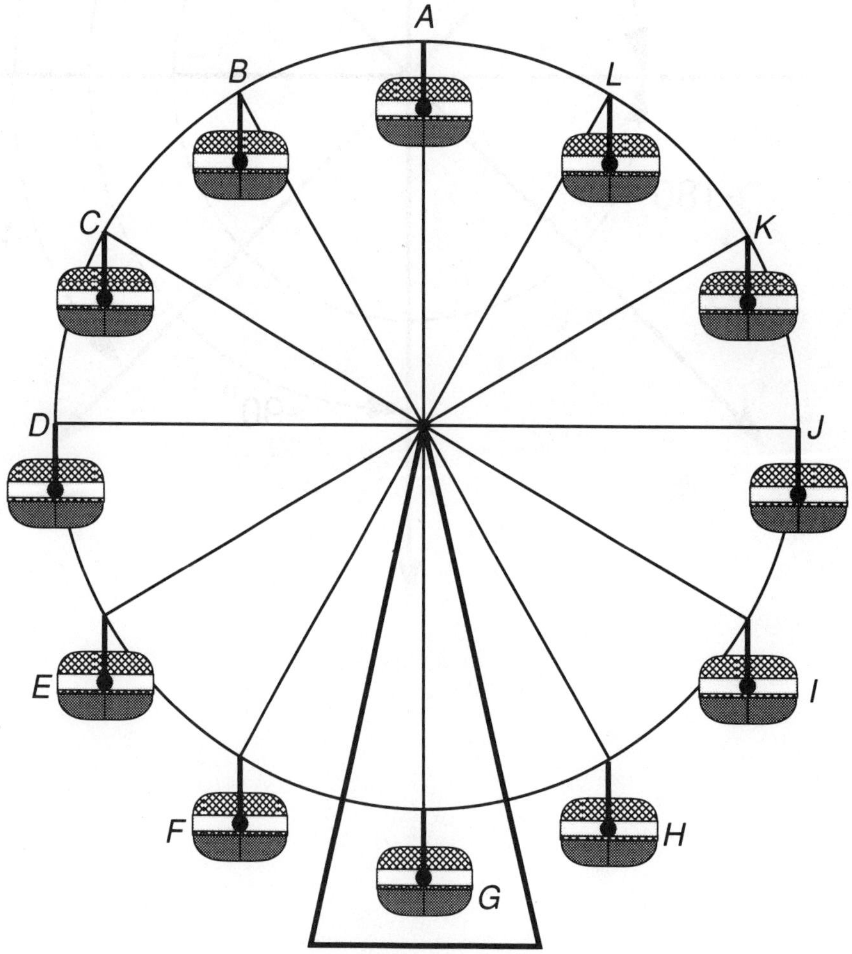

Questions

4. $3.28 = A + -5$

a. $3.28 + 5 = A + -5 + 5$ ________________

b. $3.28 + 5 = A + 0$ ________________

c. $3.28 + 5 = A$ ________________

d. $8.28 = A$ ________________

19. $(5 + 7) + -5$

Step 1 $= 5 + (7 + -5)$ ________________

Step 2 $= 5 + (-5 + -7)$ ________________

Step 3 $= (5 + -5) + 7$ ________________

Step 4 $= 0 + 7$ ________________

Step 5 $= 7$ ________________

Classification of Polygons

Number of sides	Number of vertices	Type of polygon	Example
3	3	triangle	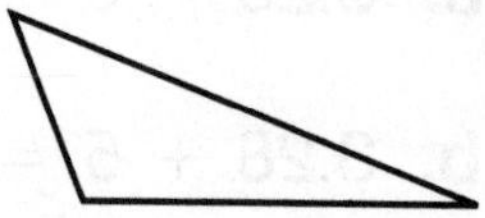
4	4	quadrilateral	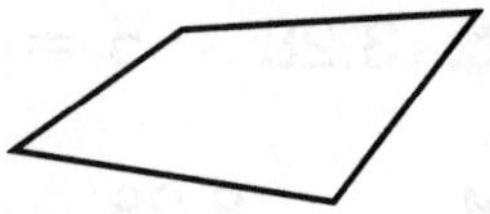
5	5	pentagon	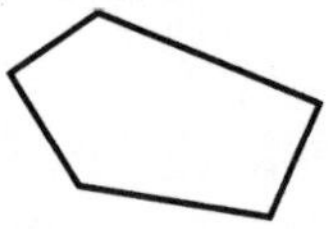
6	6	hexagon	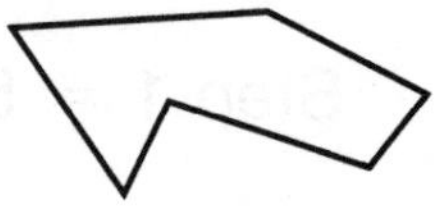
7	7	heptagon	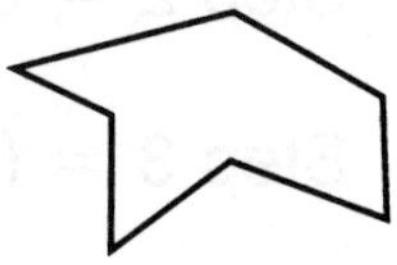
8	8	octagon	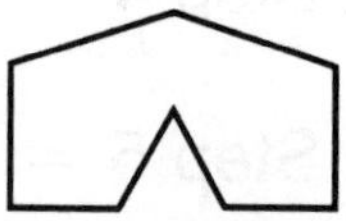
9	9	nonagon	
10	10	decagon	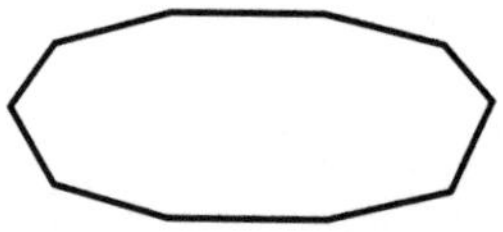

Additional Example 2

How many triangles are in this drawing?

Cost of a Phone Call

What is the cost of a phone call if the first minute costs $.25 and each additional minute costs $.18?

number of minutes talked	cost of phone call
1	
2	
3	
4	
5	
⋮	
60	
⋮	
m	

Additional Example

In a concert hall, there are 25 seats in the first row, then an additional seat on each end of each succeeding row. So the second row has 27 seats, the third row 29 seats, and so on.
How many seats are in the 6th row?

How many seats are in the 25th row?

How many seats are in the nth row?

Question 19

Integer	Square	Integer	Square
1	1	11	121
2	4	12	144
3	9	13	169
4	16	14	196
5	25	15	225
6	36	16	256
7	49	17	289
8	64	18	324
9	81	19	361
10	100	20	400

Additional Examples

3. $\angle XYZ$ is a right angle. If $m\angle WYZ = 38°$, find $m\angle WYX$.

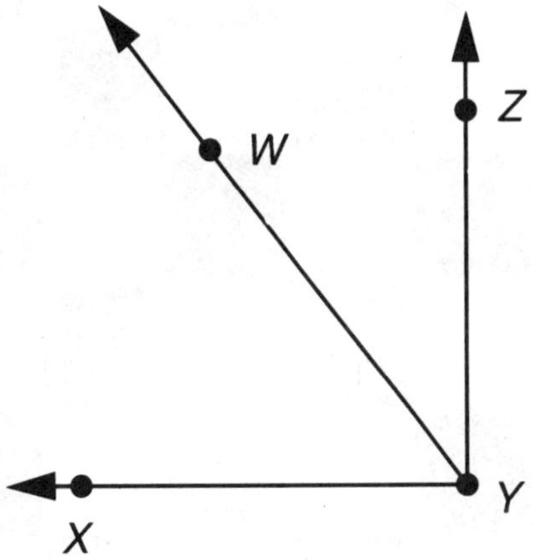

4. A square vegetable garden is 10 feet on a side. One corner, a square 5 feet on a side, will be planted with tomatoes. How many square feet are left for other vegetables?

Dominoes and a Checkerboard

Population Reference Table

Moving from state to state

Here is a state-by-state breakdown of the resident population, July 1, 1987, of each state and its gain or loss since 1980, through estimated net migration (moving, births, and deaths).

State	'87 population	gain (loss)
Alabama	4,083,000	+16,000
Alaska	523,000	+53,000
Arizona	3,386,000	+437,000
Arkansas	2,388,000	+15,000
California	27,663,000	+2,167,000
Colorado	3,296,000	+100,000
Connecticut	3,211,000	−1,000
Delaware	644,000	+19,000
D.C.	622,000	−36,000
Florida	12,023,000	+2,007,000
Georgia	6,222,000	+424,000
Hawaii	1,083,000	+23,000
Idaho	998,000	−27,000
Illinois	11,582,000	−420,000
Indiana	5,531,000	−207,000
Iowa	2,834,000	−194,000
Kansas	2,476,000	−20,000
Kentucky	3,727,000	−84,000
Louisiana	4,461,000	−74,000
Maine	1,187,000	+20,000
Maryland	4,535,000	+103,000
Massachusetts	5,855,000	−49,000
Michigan	9,200,000	−503,000
Minnesota	4,246,000	−70,000
Mississippi	2,625,000	−45,000
Missouri	5,103,000	−7,000

State	'87 population	gain (loss)
Montana	809,000	−29,000
Nebraska	1,594,000	−58,000
Nevada	1,007,000	+149,000
New Hampshire	1,057,000	+90,000
New Jersey	7,672,000	+75,000
New Mexico	1,500,000	+68,000
New York	17,825,000	−324,000
North Carolina	6,413,000	+275,000
North Dakota	672,000	−27,000
Ohio	10,784,000	−478,000
Oklahoma	3,272,000	+67,000
Oregon	2,724,000	−39,000
Pennsylvania	11,936,000	−201,000
Rhode Island	986,000	+15,000
South Carolina	3,425,000	+121,000
South Dakota	709,000	−24,000
Tennessee	4,855,000	+87,000
Texas	16,789,000	+1,245,000
Utah	1,680,000	−2,000
Vermont	548,000	+12,000
Virginia	5,904,000	+274,000
Washington	4,538,000	+145,00
West Virginia	1,897,000	−99,000
Wisconsin	4,807,000	−133,000
Wyoming	490,000	−28,000

Question 24

$(b - a) + (a - b)$

Step 1: $= (b + -a) + (a + -b)$ ________________

Step 2: $= b + (-a + a) + -b$ ________________

Step 3: $= b + 0 + -b$ ________________

Step 4: $= b + -b$ ________________

Step 5: $= 0$ ________________

Question 19

Give a reason for each step in this solution for *x*.

$$a - x = b$$

Step 1: $a + -x = b$ ____________________

Step 2: $(a + -x) + x = b + x$ ____________________

Step 3: $a + (-x + x) = b + x$ ____________________

Step 4: $a + 0 = b + x$ ____________________

Step 5: $a = b + x$ ____________________

Step 6: $-b + a = -b + (b + x)$ ____________________

Step 7: $-b + a = (-b + b) + x$ ____________________

Step 8: $-b + a = 0 + x$ ____________________

Step 9: $-b + a = x$ ____________________

Map of Drexel and Ellis Avenues

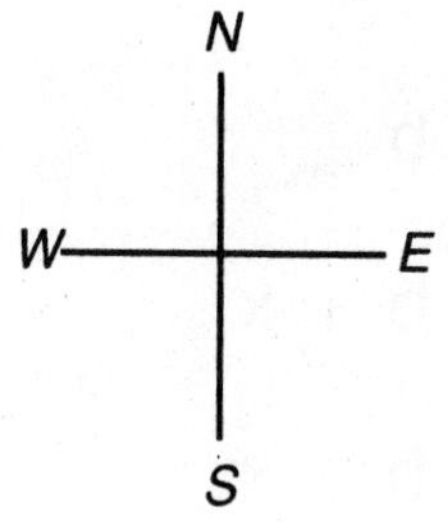

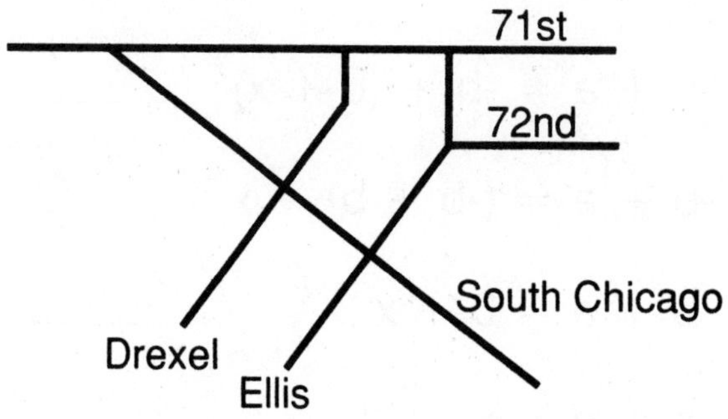

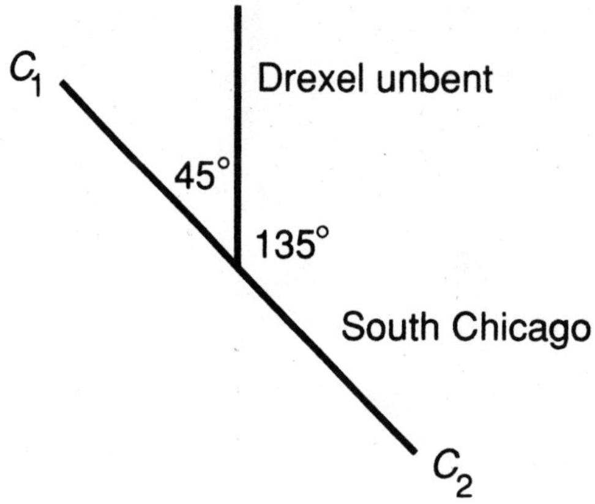

Angles and Parallel Lines

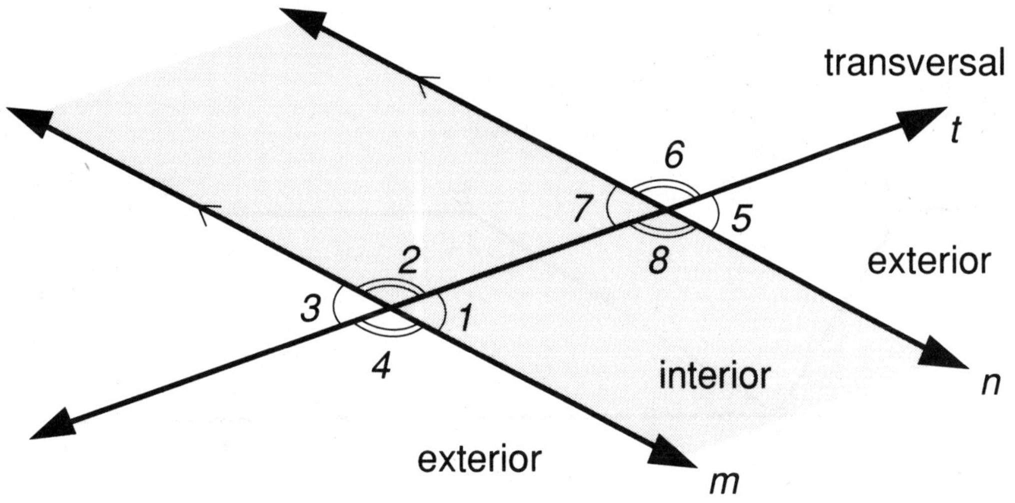

$m \parallel n$

Name	Angles in above figure	Measures
Corresponding angles	1 and 5, 2 and 6 3 and 7, 4 and 8	equal
Interior angles	1, 2, 7, 8	
Alternate interior angles	2 and 8, 1 and 7	equal
Exterior angles	3, 4, 5, 6	

Question 23

23. *ABCD* and *AEFG* are parallelograms and $\overline{AB}$ and $\overline{AE}$ are perpendicular. m∠*ABC* = 147°. *G, A, D,* and *H* are on the same line. Find the measure of: **a.**∠BAG; **b.**∠GAE; **c.**∠AGF; **d.**∠CDH.

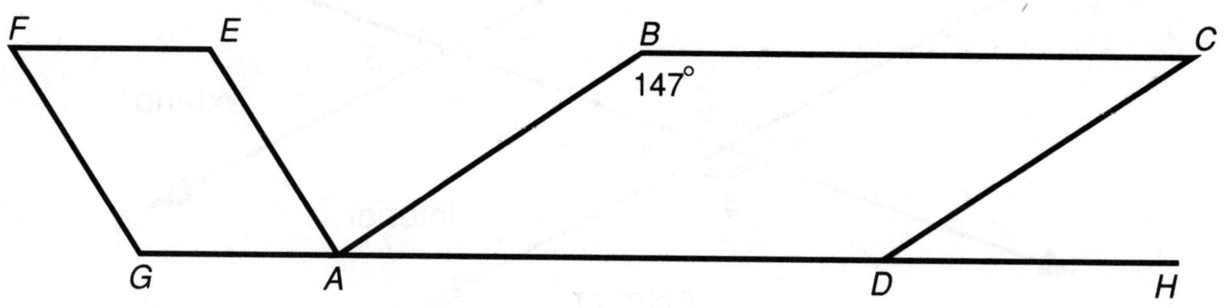

Additional Examples

Find the measure of ∠1 in each figure.

1.

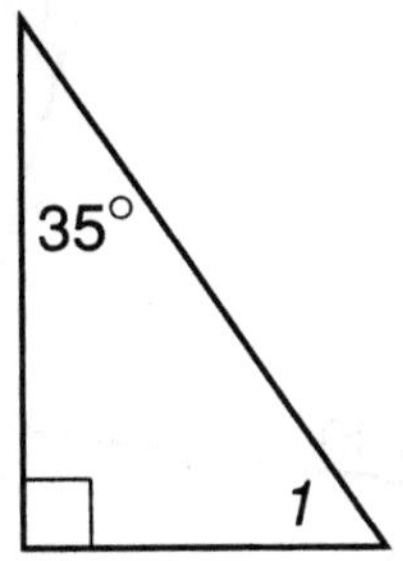

2.

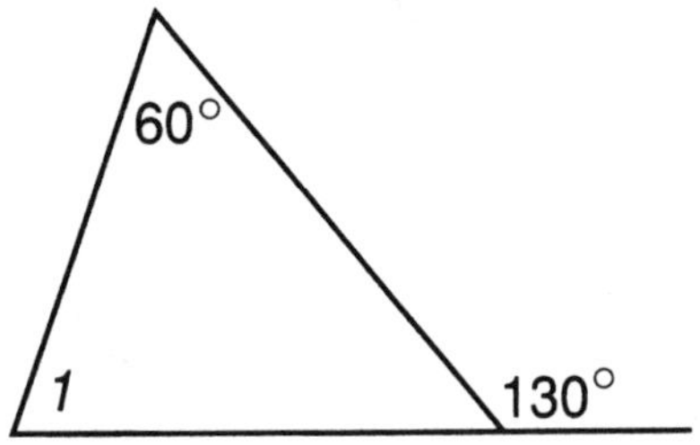

3.

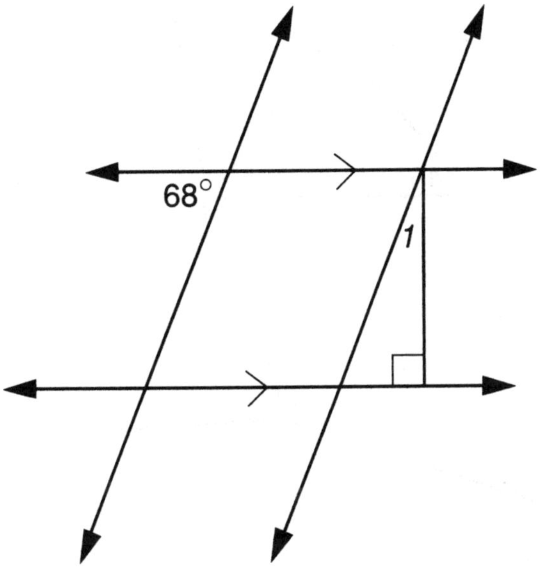

Questions

10.

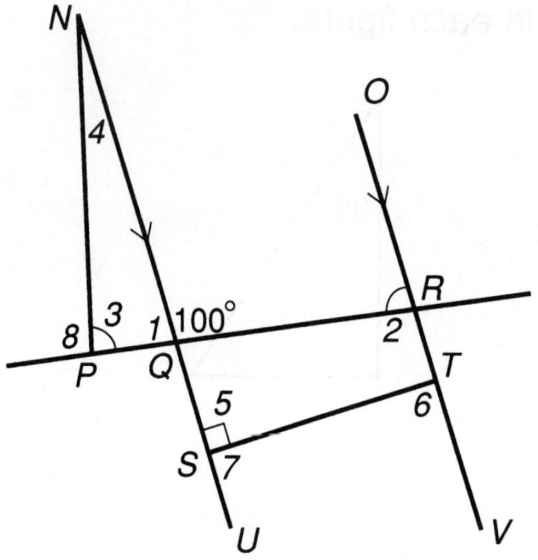

16.

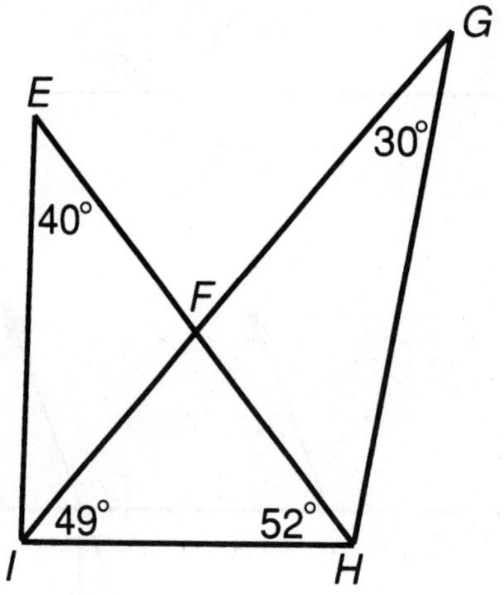

29.

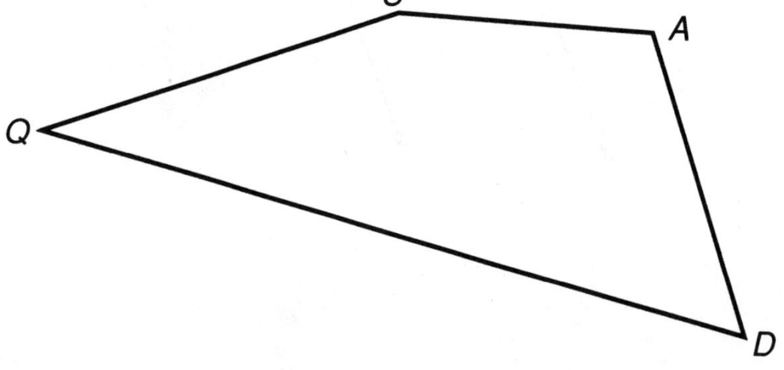

The Triangle-Sum Property

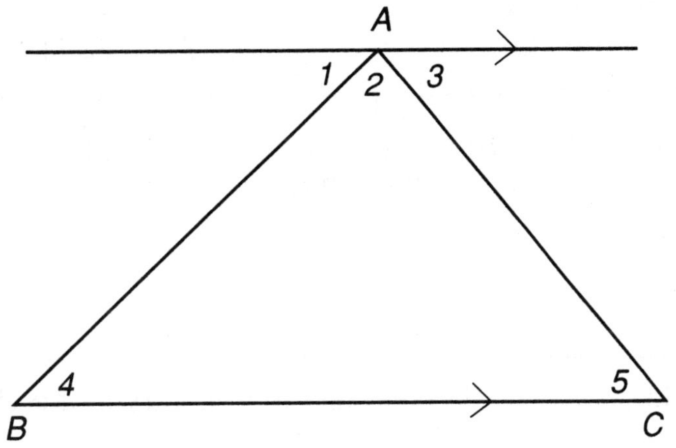

Example 1

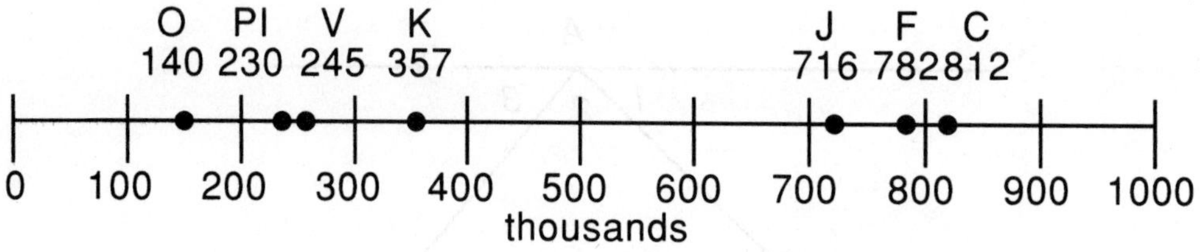

O PI V K
140 230 245 357
J F C
716 782 812
0 100 200 300 400 500 600 700 800 900 1000
thousands

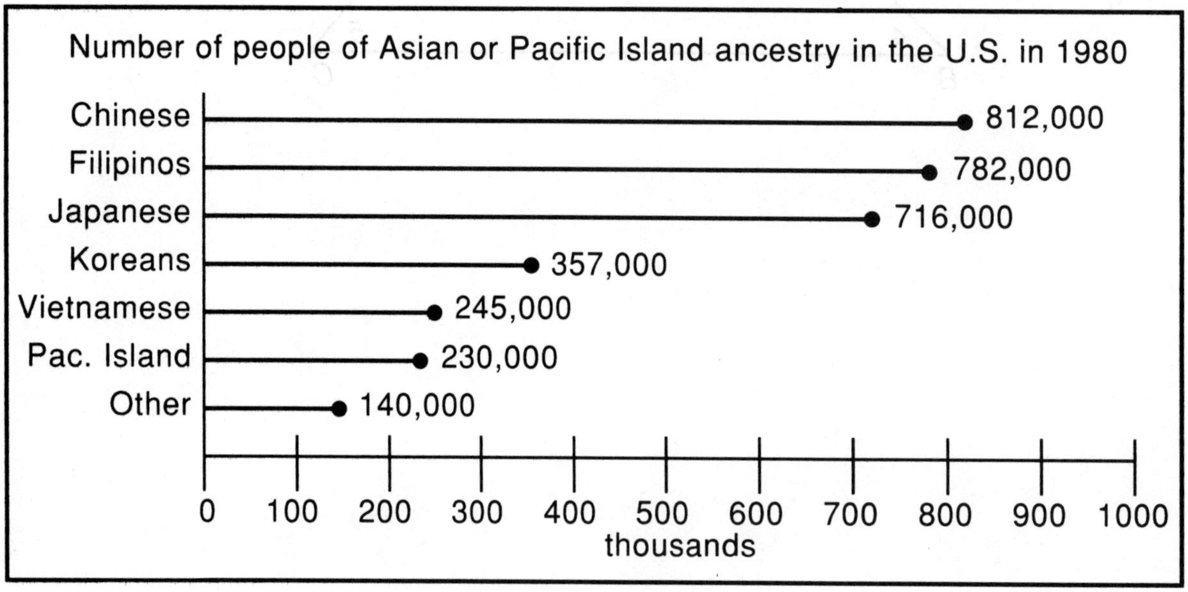

Number of people of Asian or Pacific Island ancestry in the U.S. in 1980
Chinese
812,000
Filipinos
782,000
Japanese
716,000
Koreans
357,000
Vietnamese
245,000
Pac. Island
230,000
Other
140,000
0 100 200 300 400 500 600 700 800 900 1000
thousands

Example 2

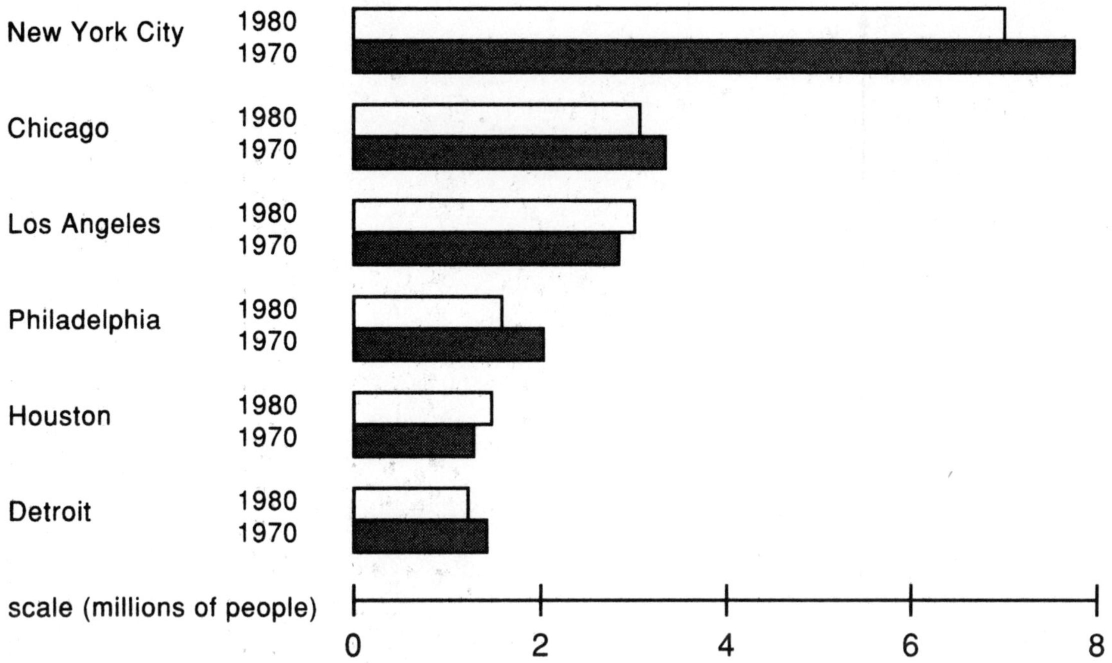
New York City
1980
1970
Chicago
1980
1970
Los Angeles
1980
1970
Philadelphia
1980
1970
Houston
1980
1970
Detroit
1980
1970
scale (millions of people)
0
2
4
6
8

Graph Paper

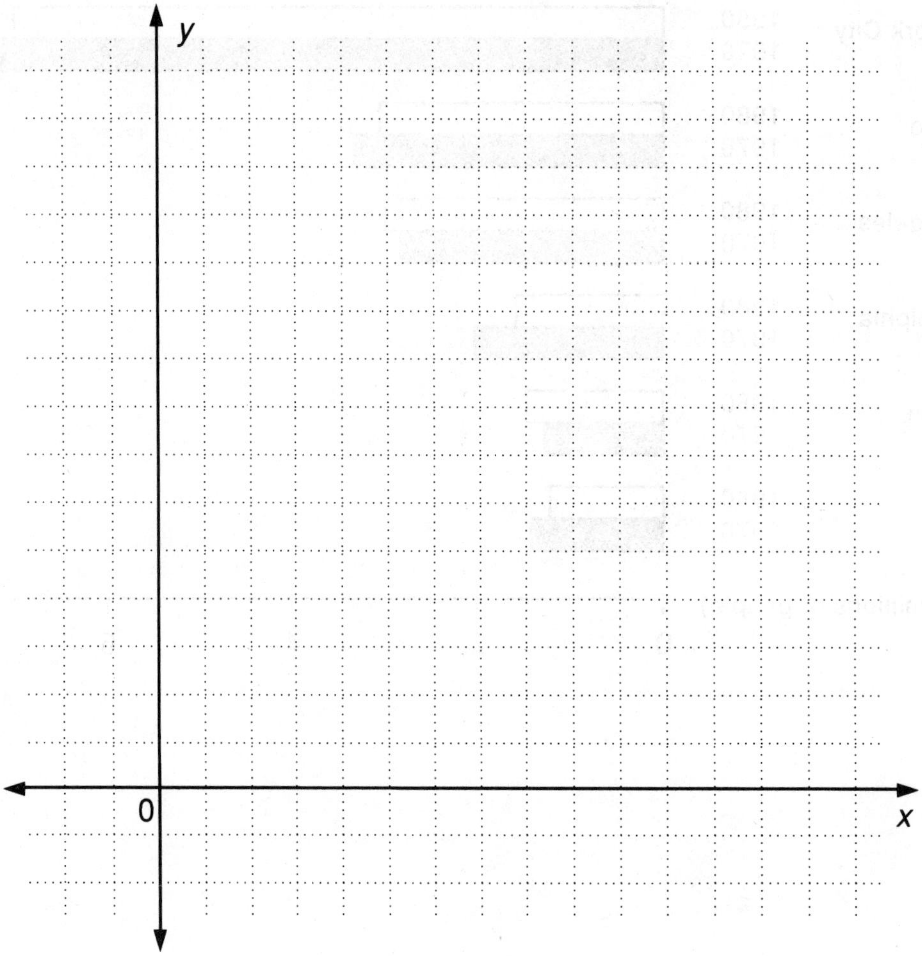

Graph Paper (Four Quadrants)

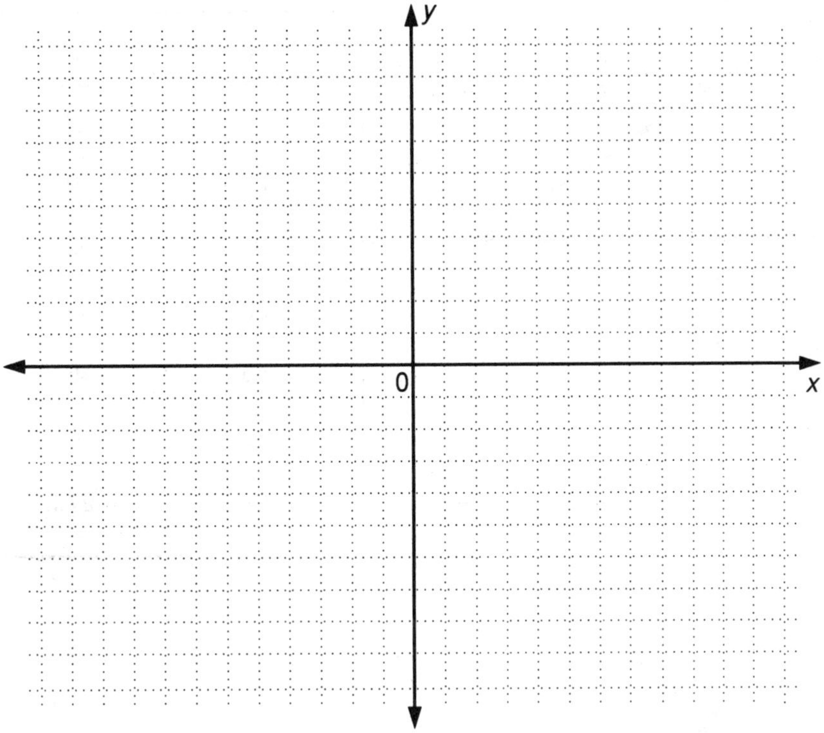

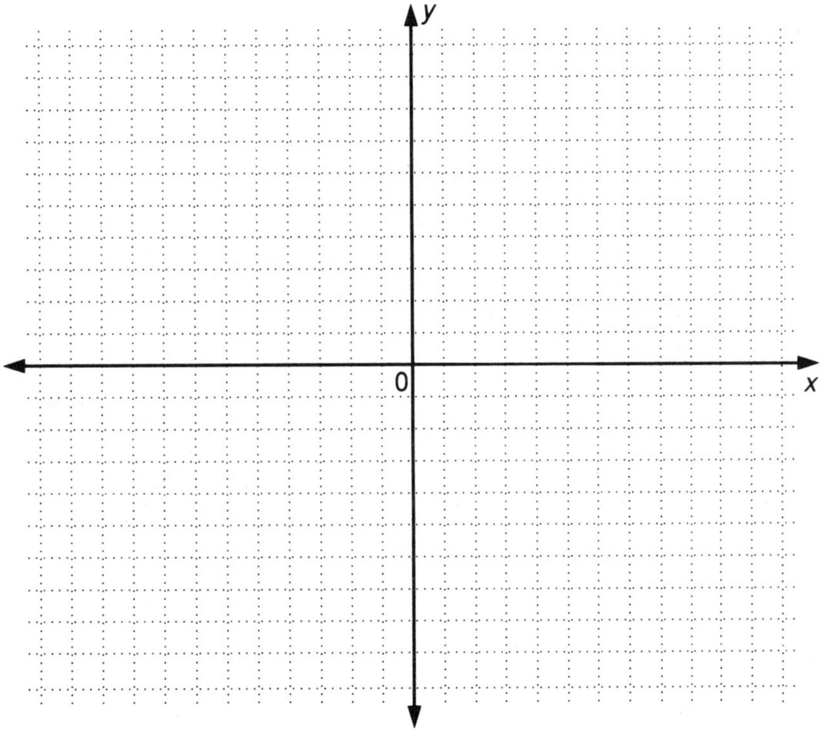

Graphing an Equation ($x + y = 90$)

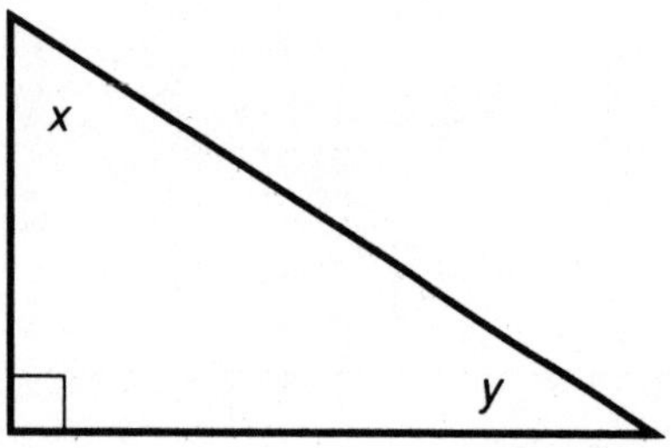

x	y
10	80
20	70
35	55
52	38
45	45
80	10
62.8	27.2

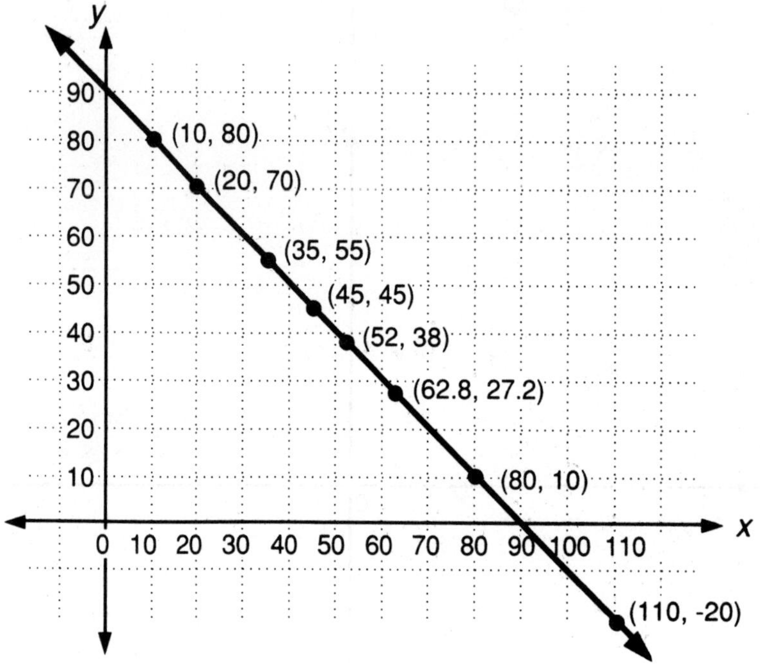

Translation Images

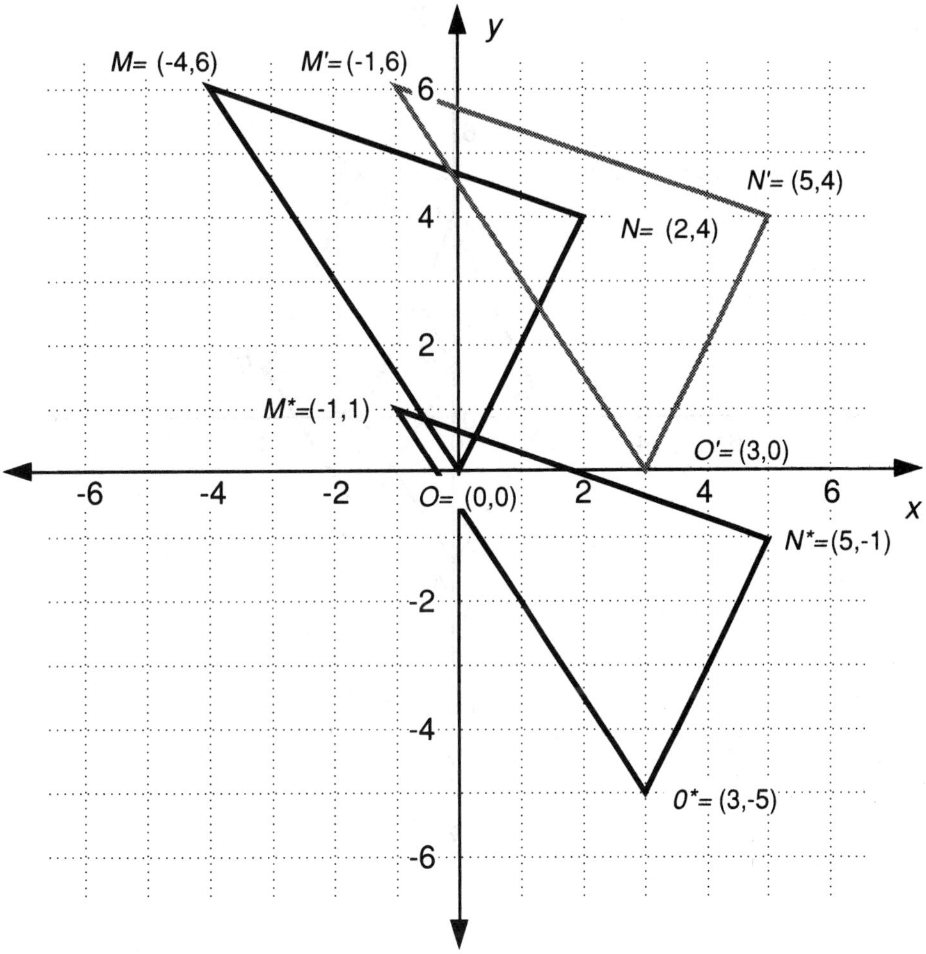

Additional Examples

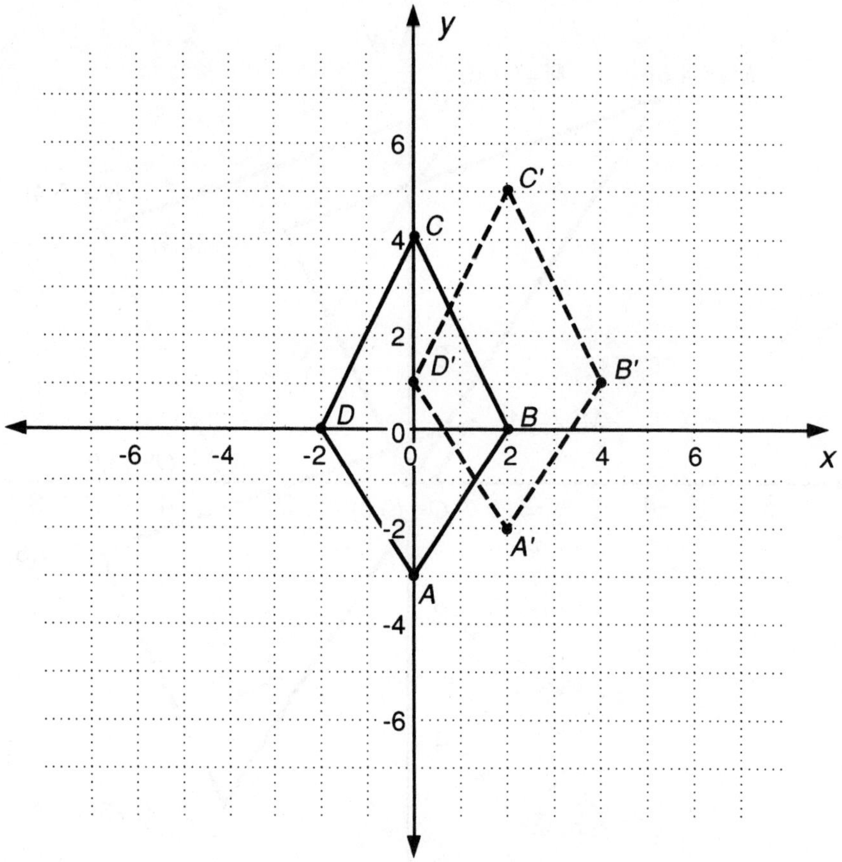

Question 16

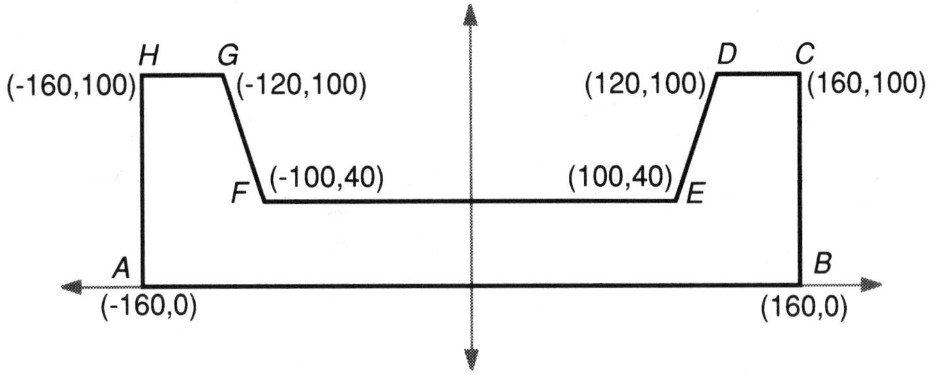
H
G
D
C
(-160,100)
(-120,100)
(120,100)
(160,100)
F
(-100,40)
(100,40)
E
A
B
(-160,0)
(160,0)

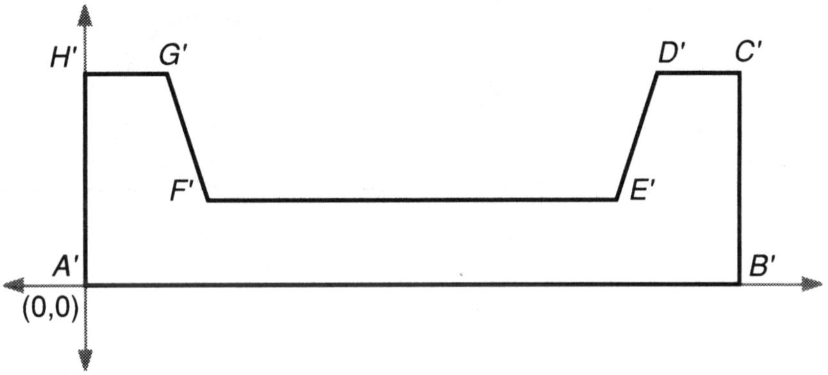
H'
G'
D'
C'
F'
E'
A'
B'
(0,0)

Additional Examples

1.

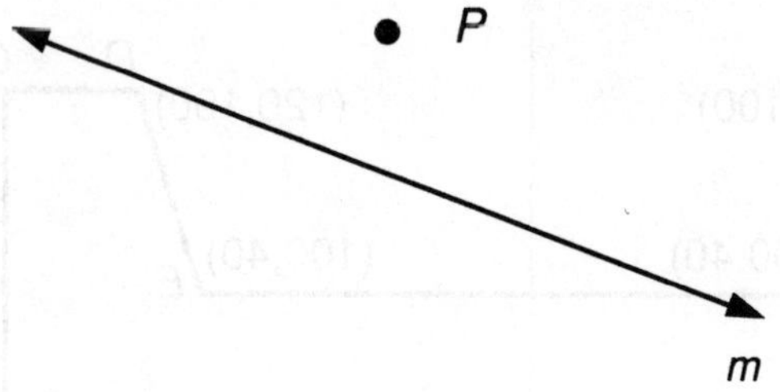

2.

3.

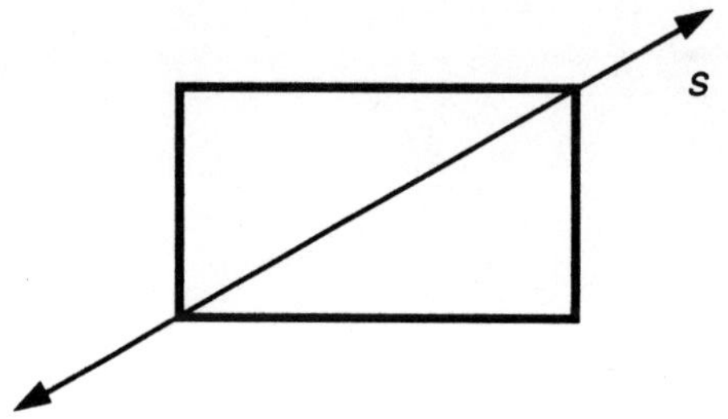

Questions 8–11

8.

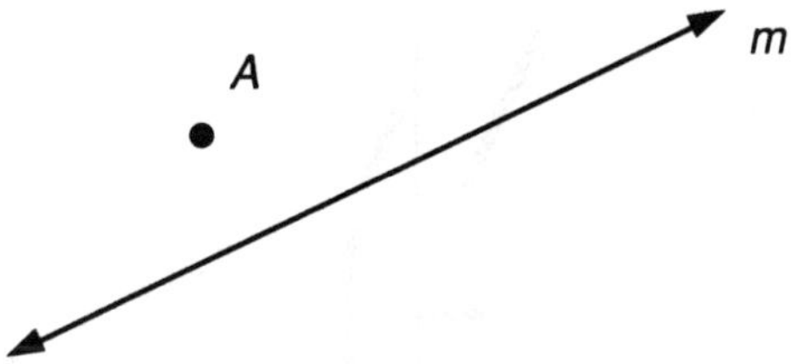

9.

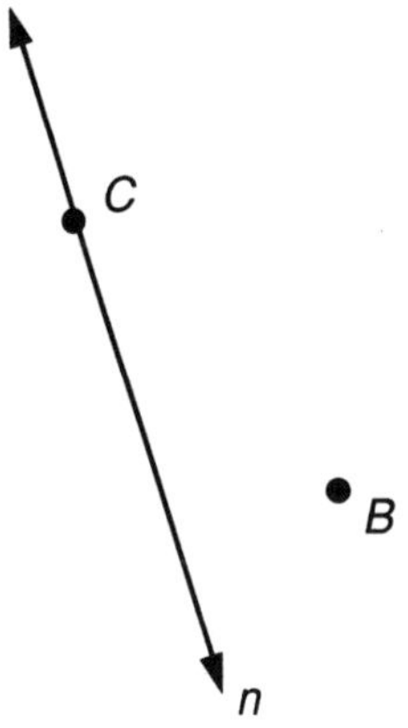

10.

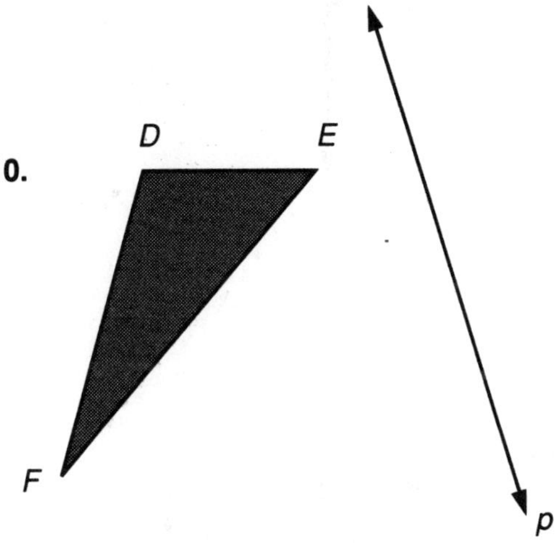

11.

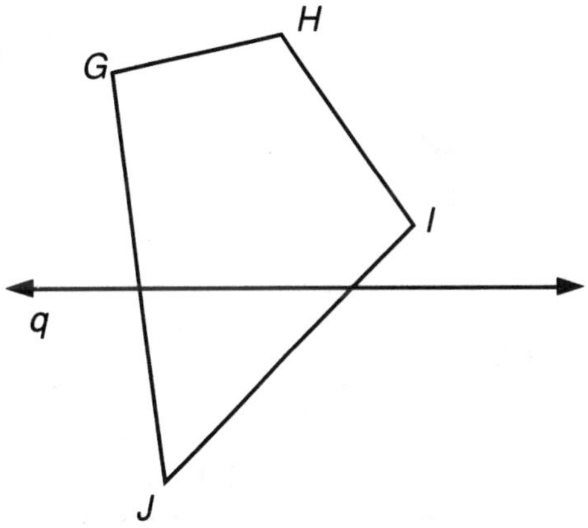

Questions 15–18

15.

16.

17.

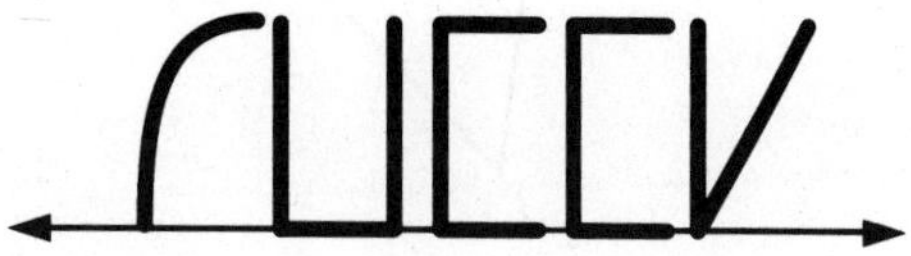

18.

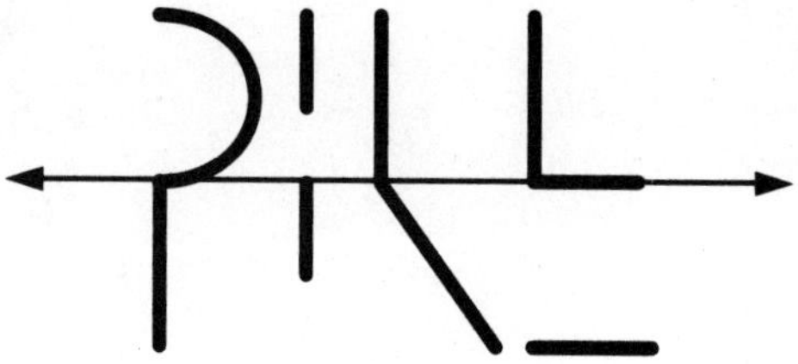

Questions 10–13

10. 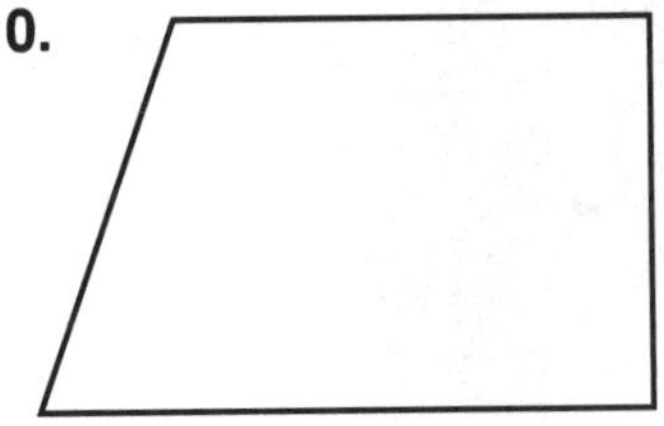

11. 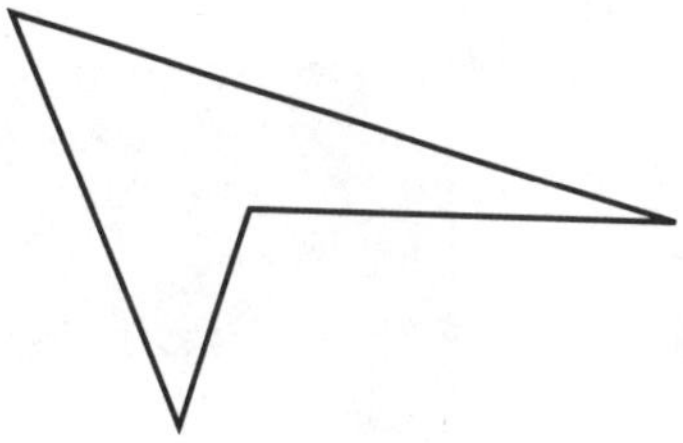

12. 

13. 

Questions 16 and 28

16.

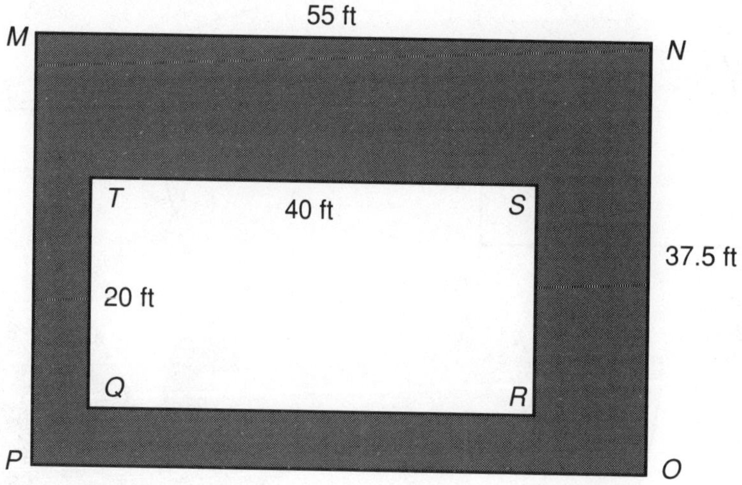

28.

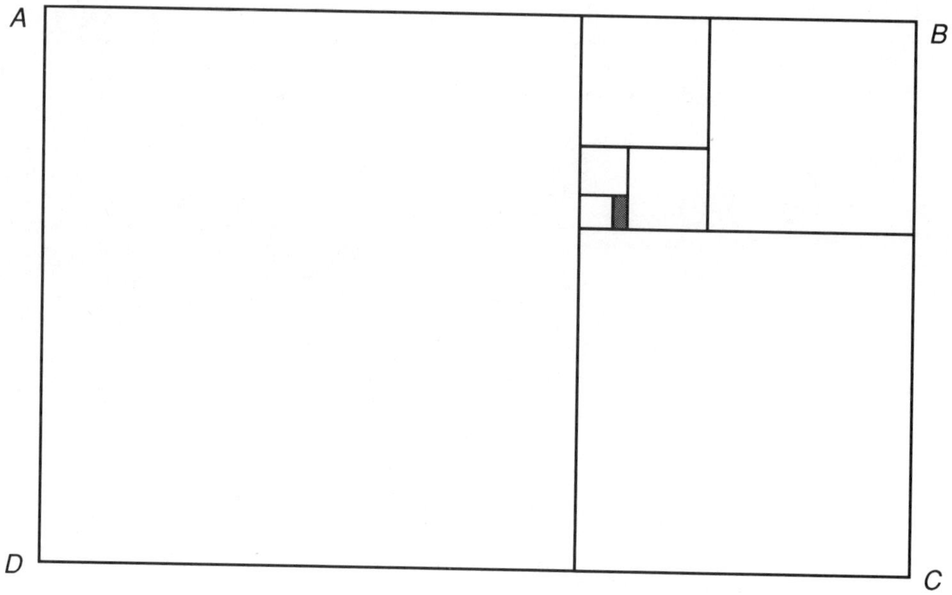

Surface Area of a Cereal Box

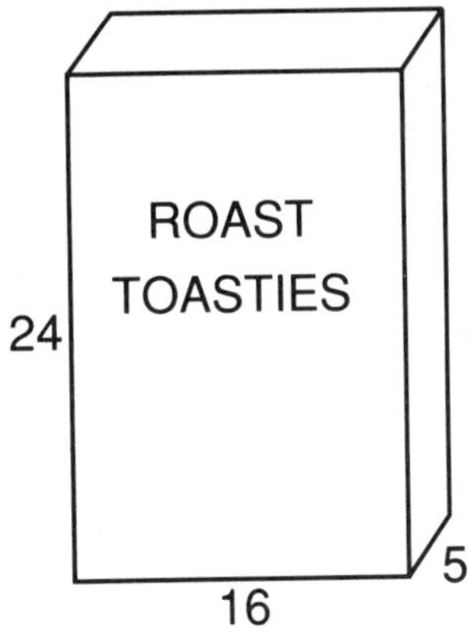

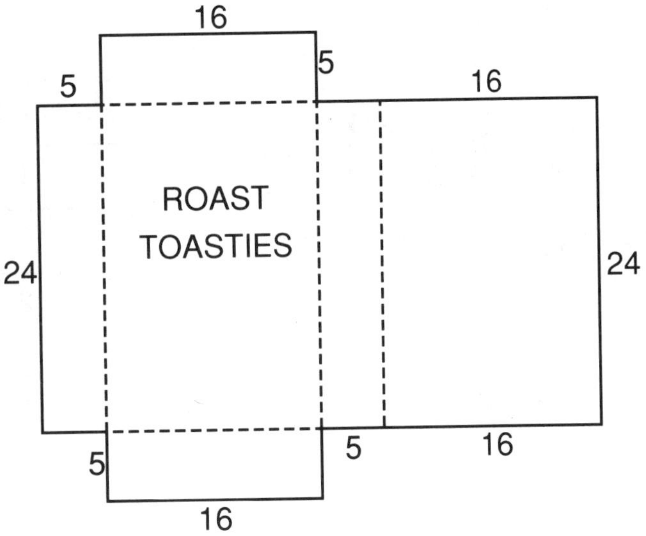

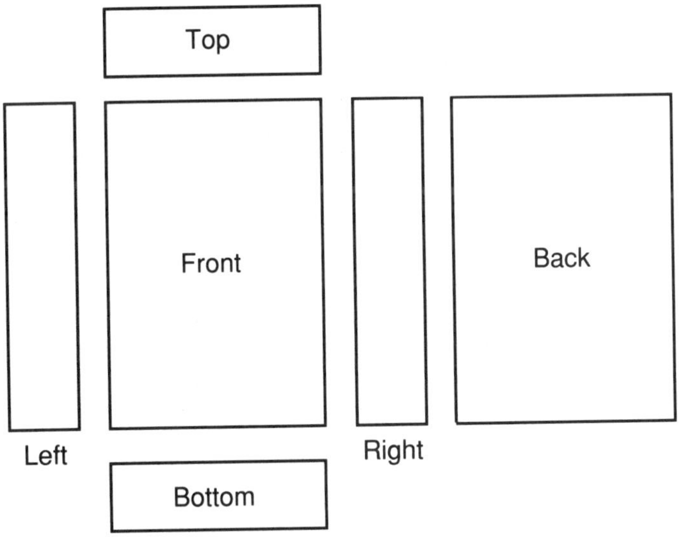

Face	Area (in cm^2)
Left	$5 \cdot 24 = 120$
Right	$5 \cdot 24 = 120$
Front	$16 \cdot 24 = 384$
Back	$16 \cdot 24 = 384$
Top	$5 \cdot 16 = 80$
Bottom	$5 \cdot 16 = 80$
TOTAL	1168 cm^2

Expansions

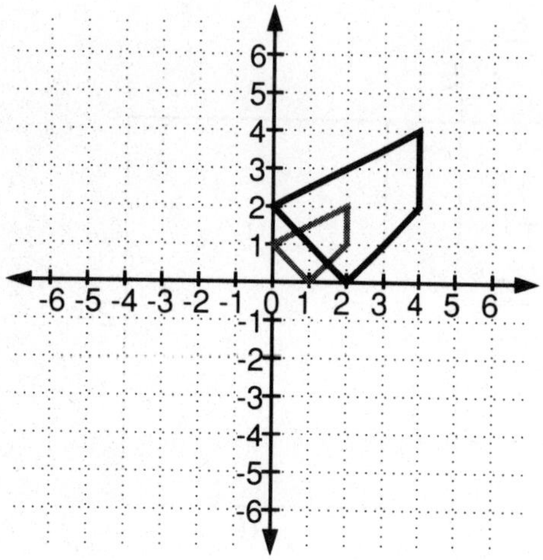

Preimage Point	Image Point
(1, 0)	(2, 0)
(0, 1)	(0, 2)
(2, 2)	(4, 4)
(2, 1)	(4, 2)

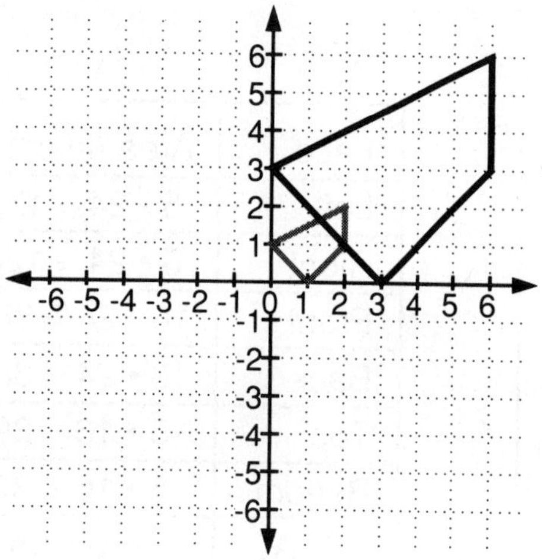

Preimage Point	Image Point
(1, 0)	(3, 0)
(0, 1)	(0, 3)
(2, 2)	(6, 6)
(2, 1)	(6, 3)

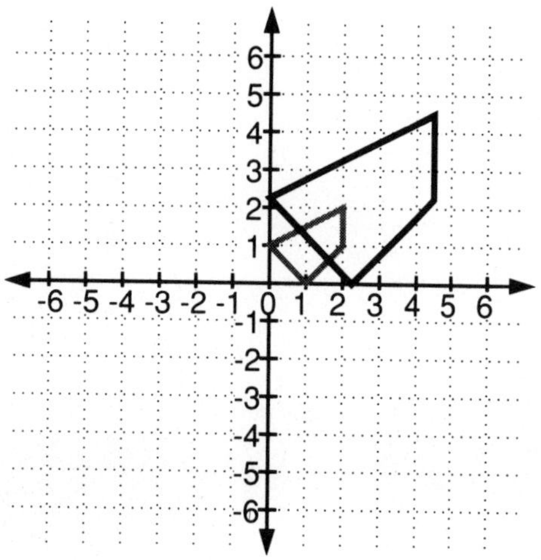

Preimage Point	Image Point
(1, 0)	(2.25, 0)
(0, 1)	(0, 2.25)
(2, 2)	(4.5, 4.5)
(2, 1)	(4.5, 2.25)

Contractions

- Graph the original figure under a size change of the given magnitude.

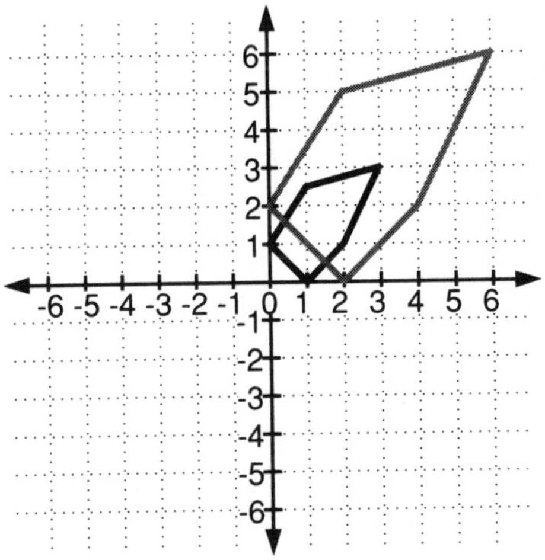

Preimage Point	Image Point
(0, 2)	(0, 1)
(2, 0)	(1, 0)
(4, 2)	(2, 1)
(6, 6)	(3, 3)
(2, 5)	(1, 2.5)

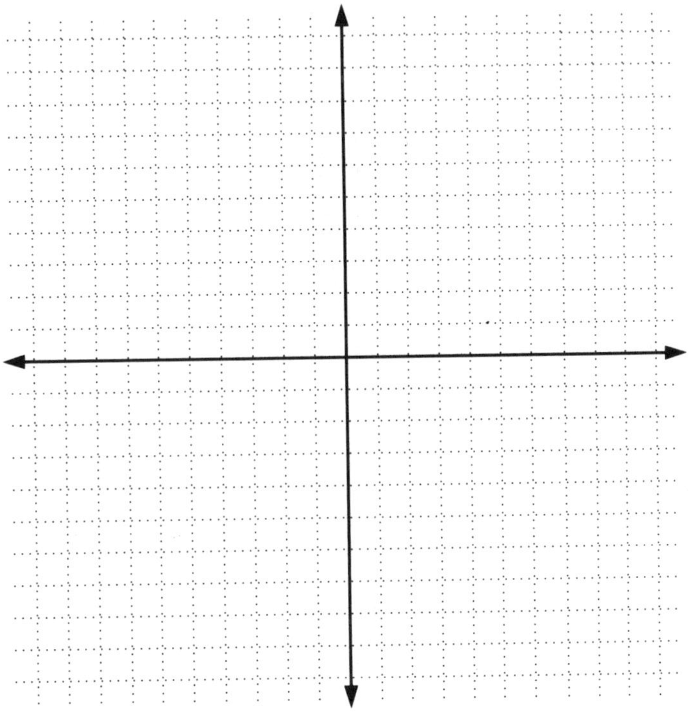

Questions

11–12.

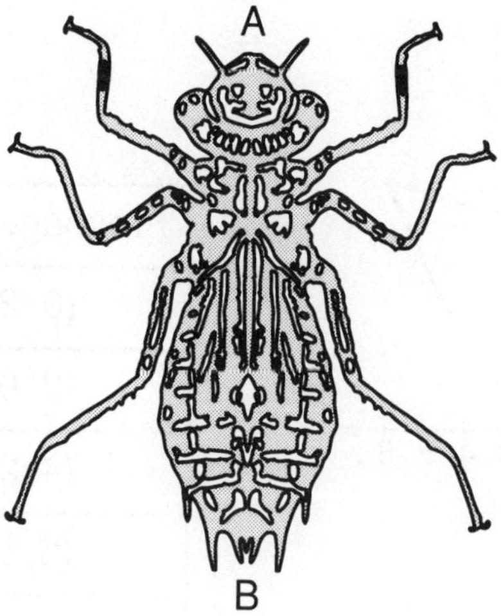

31.

Size Changes

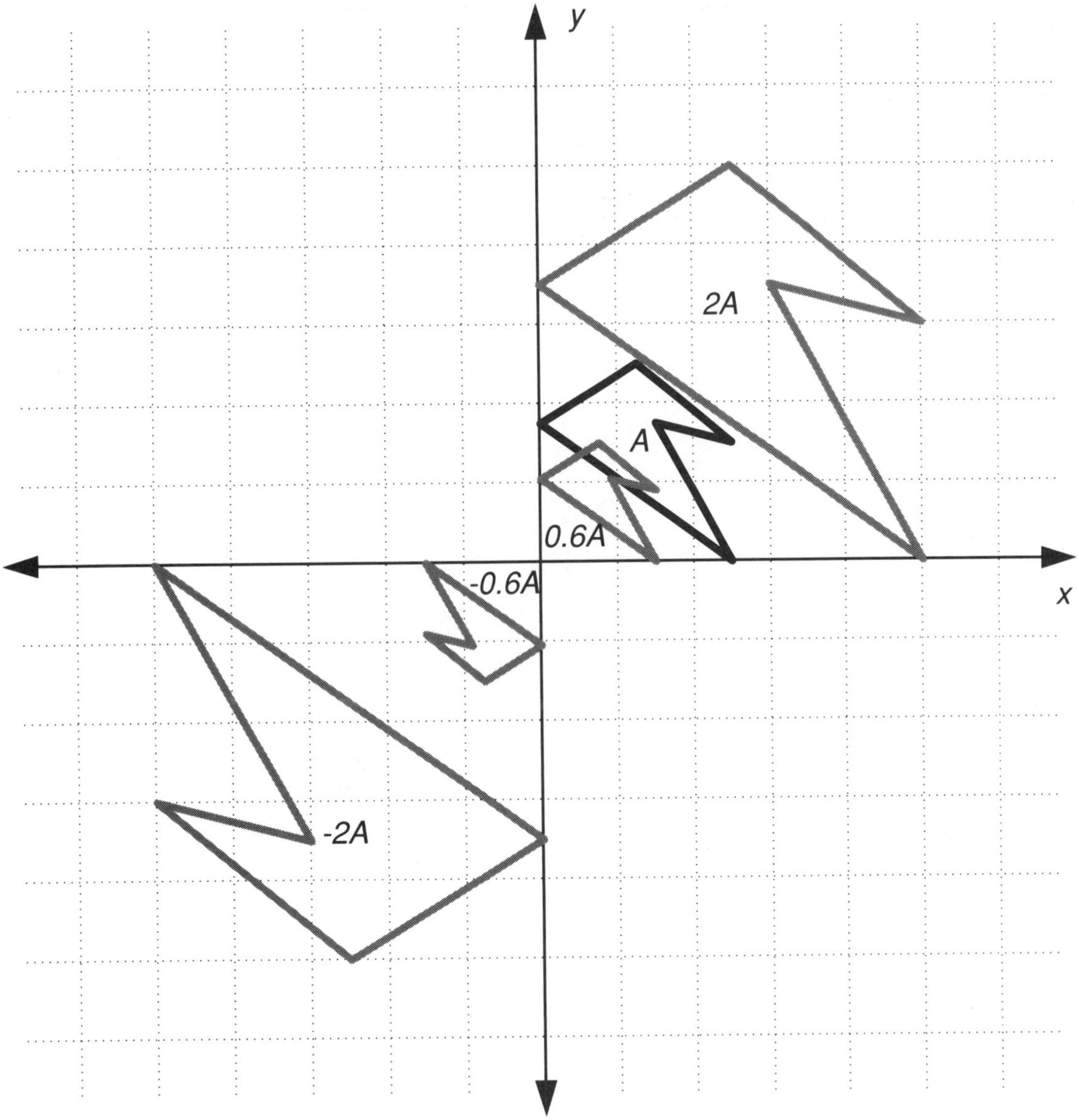

Expansions with Negative Magnitude

Expansion, Magnitude 2	
Preimage Point	Image Point
$A = (-2, 3)$	$A' = (-4, 6)$
$B = (0,-2)$	$B' = (0,-4)$
$C = (3,-2)$	$C' = (6,-4)$
$D = (3, 1)$	$D' = (6, 2)$

Expansion, Magnitude -2	
Preimage Point	Image Point
$A = (-2, 3)$	$A'' = (4,-6)$
$B = (0,-2)$	$B'' = (0, 4)$
$C = (3,-2)$	$C'' = (-6, 4)$
$D = (3, 1)$	$D'' = (-6,-2)$

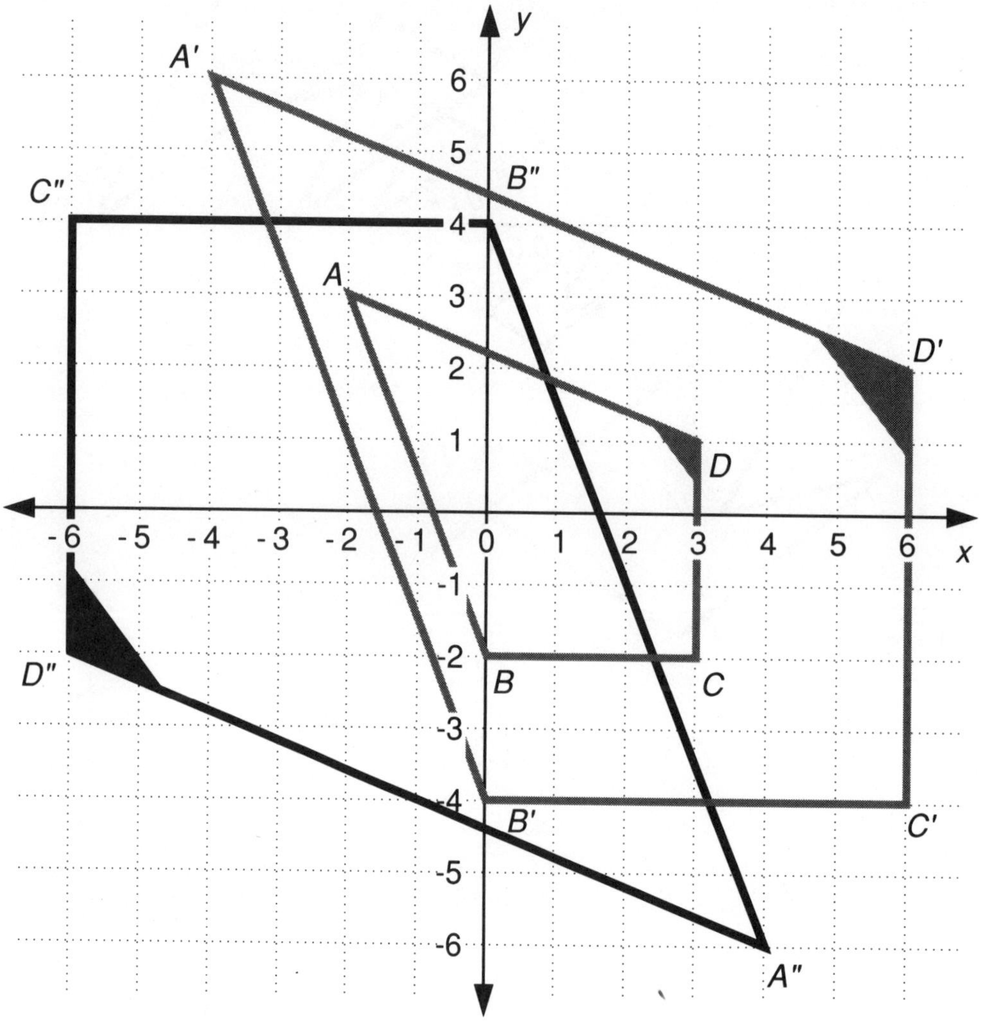

Question 24

Apply a size change of magnitude -1 to $\triangle ABC$. Call the image $\triangle A'B'C'$. How do $\triangle ABC$ and $\triangle A'B'C'$ compare?

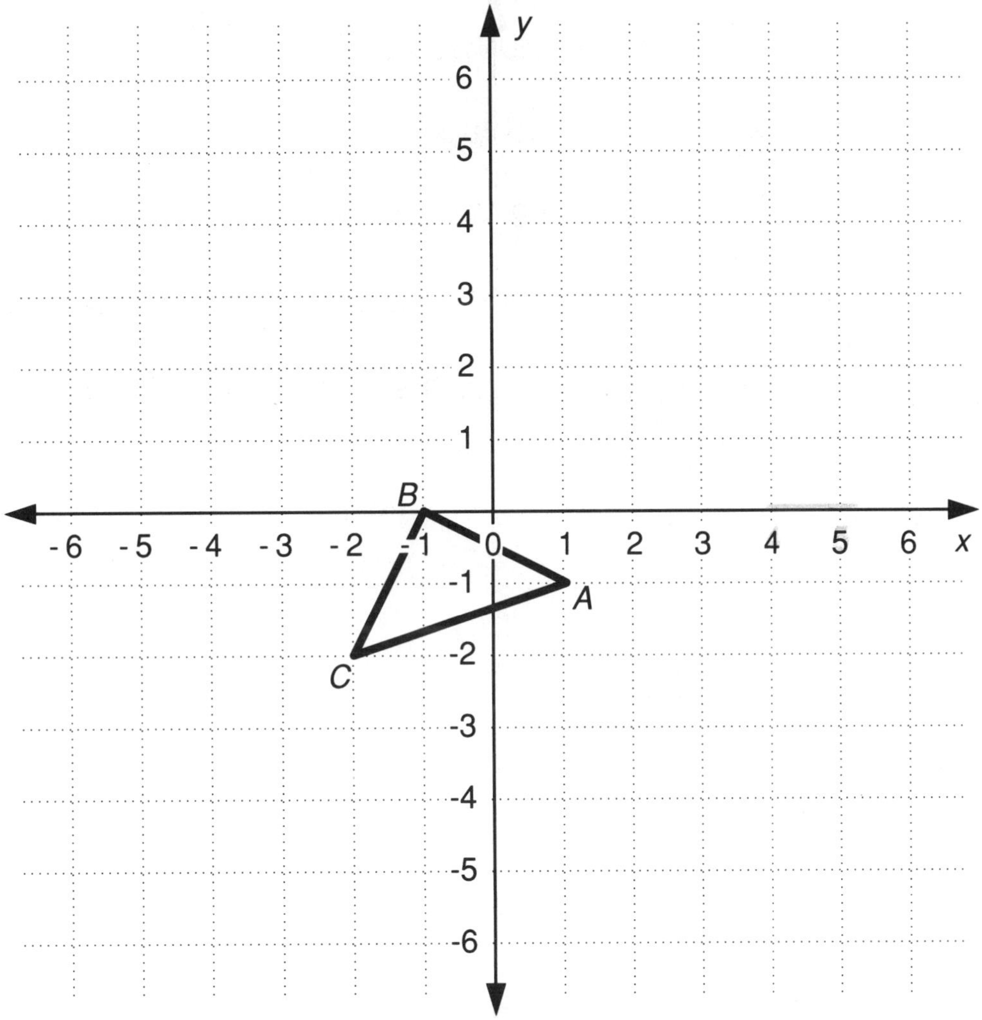

Question 23

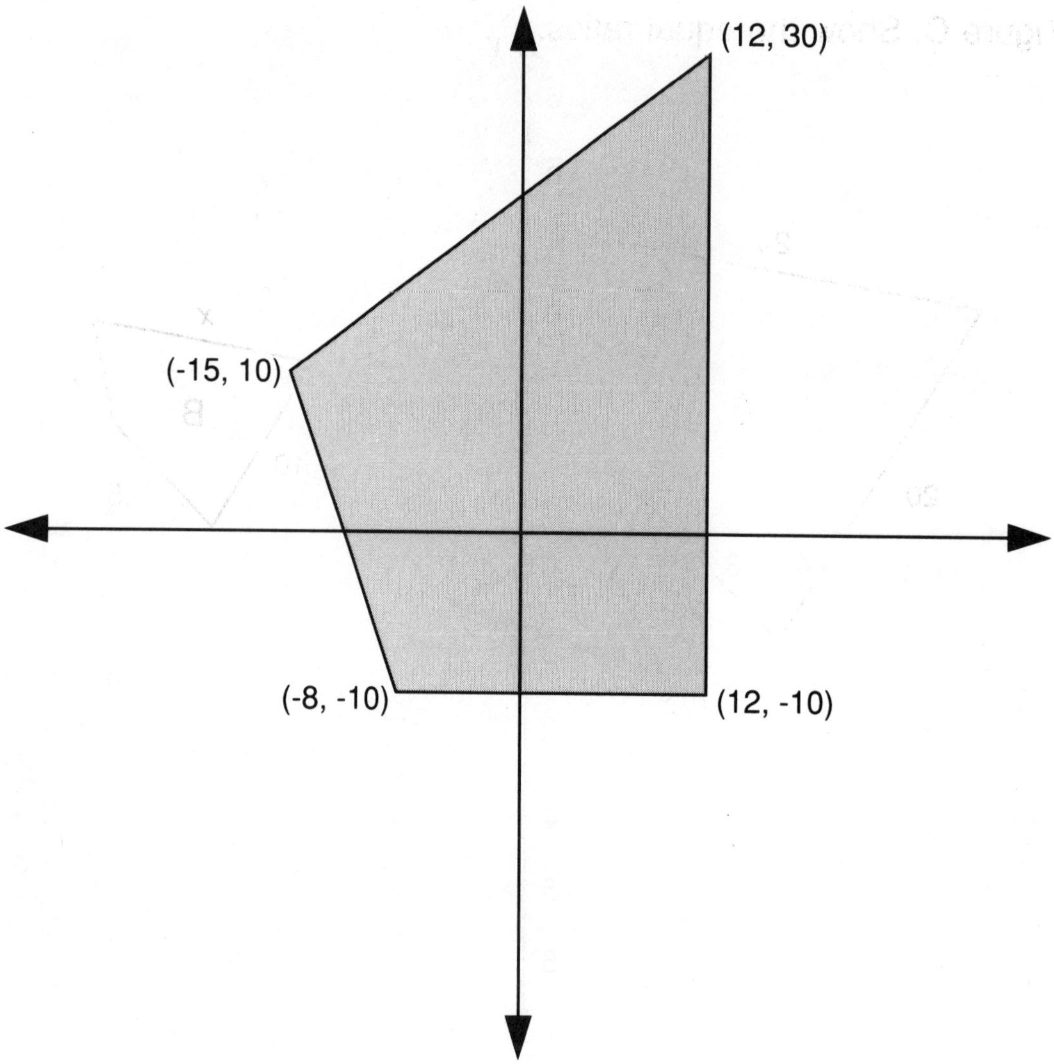

Similar Figures

What would be the dimensions of Figure C if that figure were the image of Figure A under an expansion of the given magnitude?

Draw Figure C. Show the equal ratios.

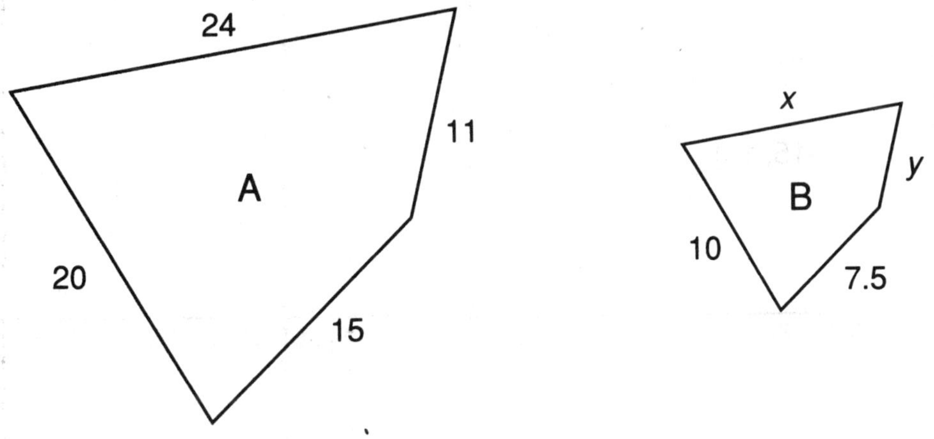

Additional Examples

1. Given $\triangle RUN$ is similar to $\triangle HOP$, list the pairs of corresponding sides.

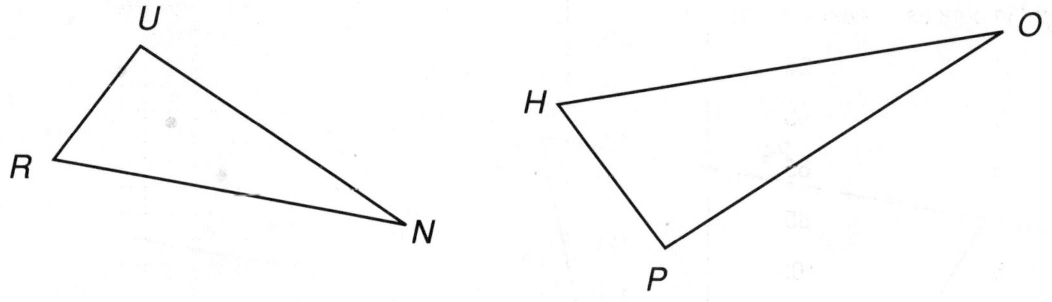

2. These two trapezoids are similar. Find the missing sides.

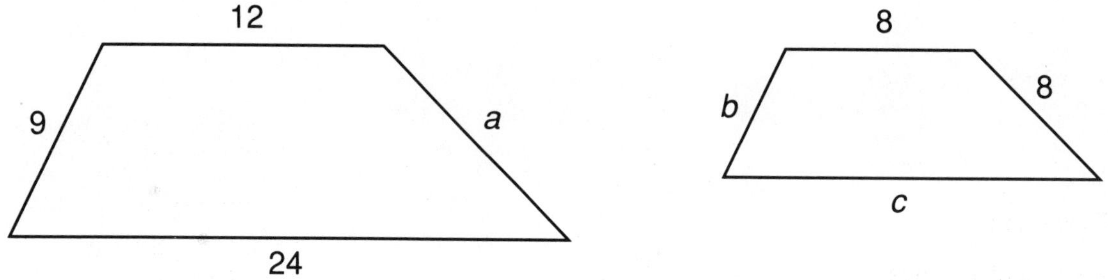

Displaying Postal Rates

weight in ounces	cost in cents
1	25
2	45
3	65
4	85
5	105
6	125
7	145
8	165
9	185
10	205

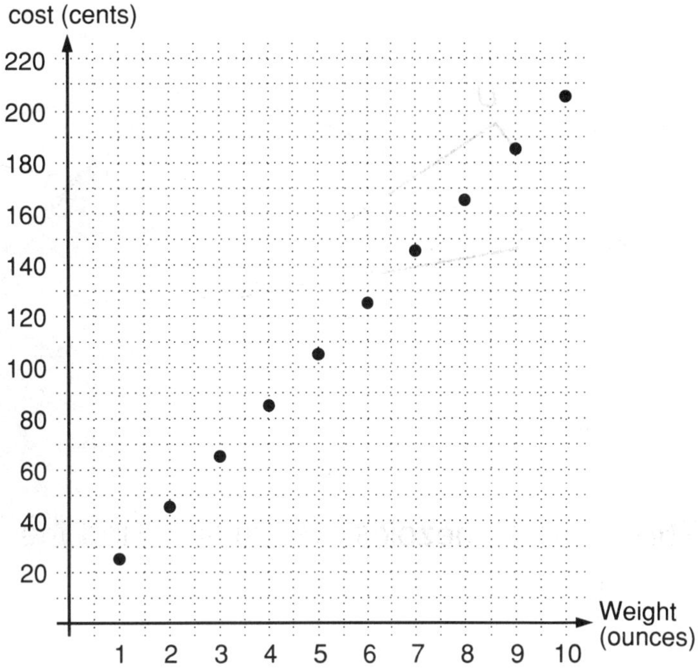

Postal Rates

w = weight in ounces.

c = cost in pennies.

1965: $c = 5w$.

1975: $c = 11w + 2$.

1988: $c = 20w + 5$.

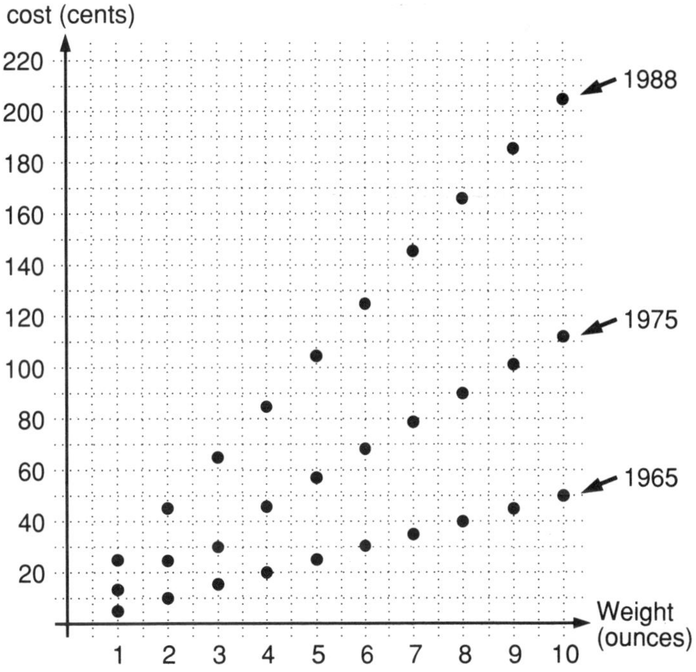

Area of a Lune and a Right Triangle

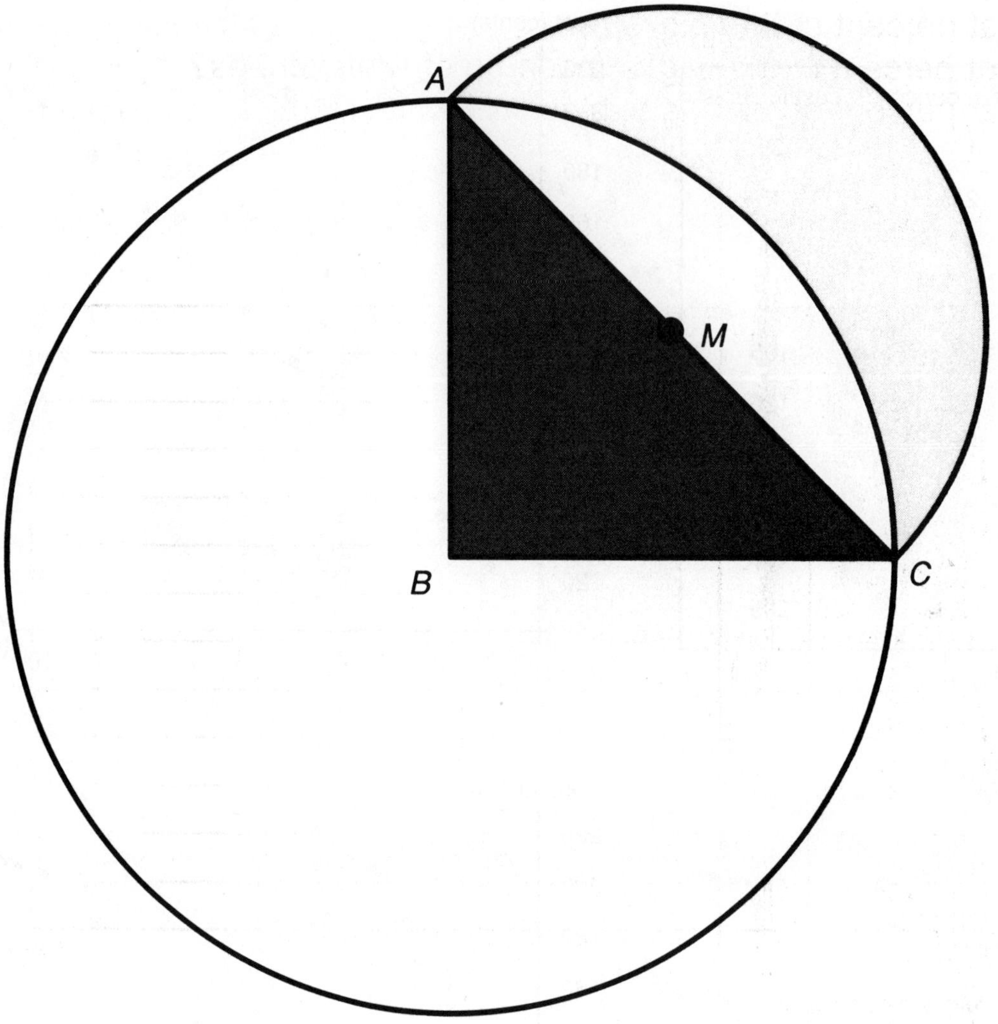

Question 26

The United States flag has 7 red stripes and 6 white ones. The stripes all have the same thickness. The upper left corner is blue with 50 white stars. Use the given dimensions of the flag to determine:

a. What percent of the flag is red?
b. What percent of the flag is made up of white stripes?
c. What percent of the flag is blue with white stars?

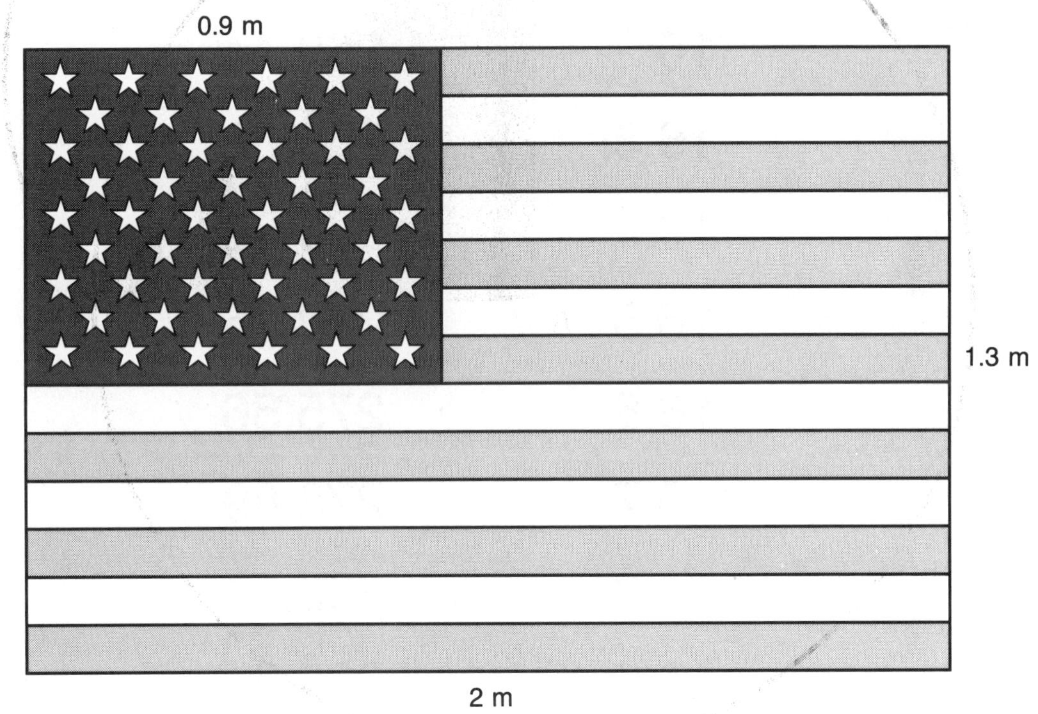

Questions 27–30

Number	Square
16.0	256.00
16.1	259.21
16.2	262.44
16.3	265.69
16.4	268.96
16.5	272.25
16.6	275.56
16.7	278.89
16.8	282.24
16.9	285.61
17.0	289.00

Proof of the Pythagorean Theorem

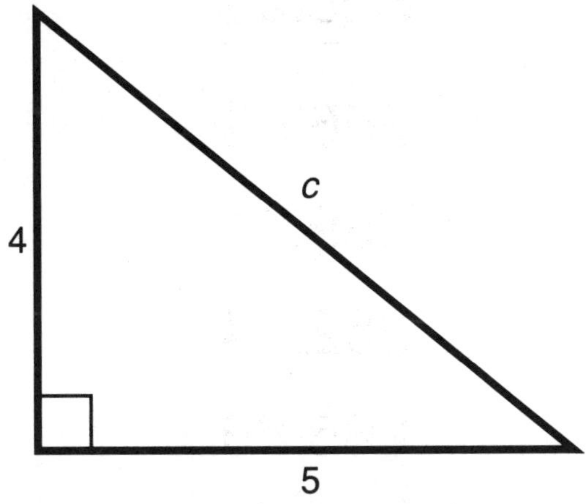

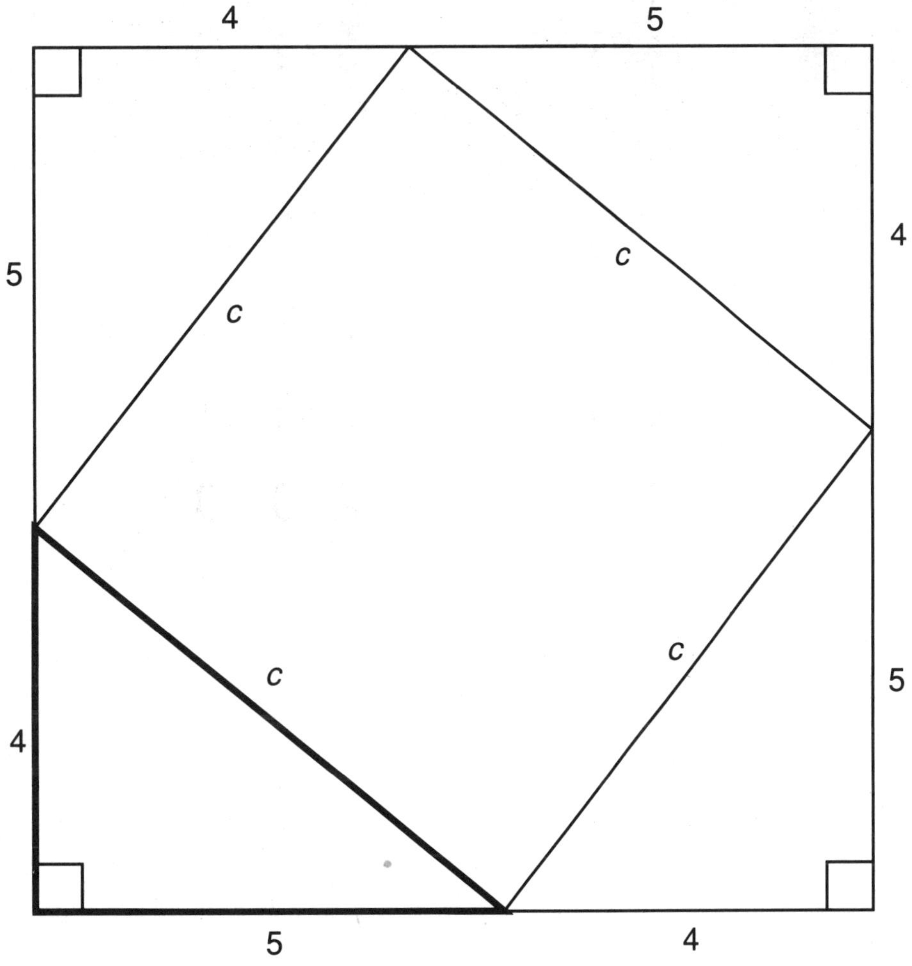

Question 24

The Pythagoreans worshipped numbers and loved music. They discovered relationships between the lengths of strings and the musical tones they give. What are some of the relationships the Pythagoreans discovered?

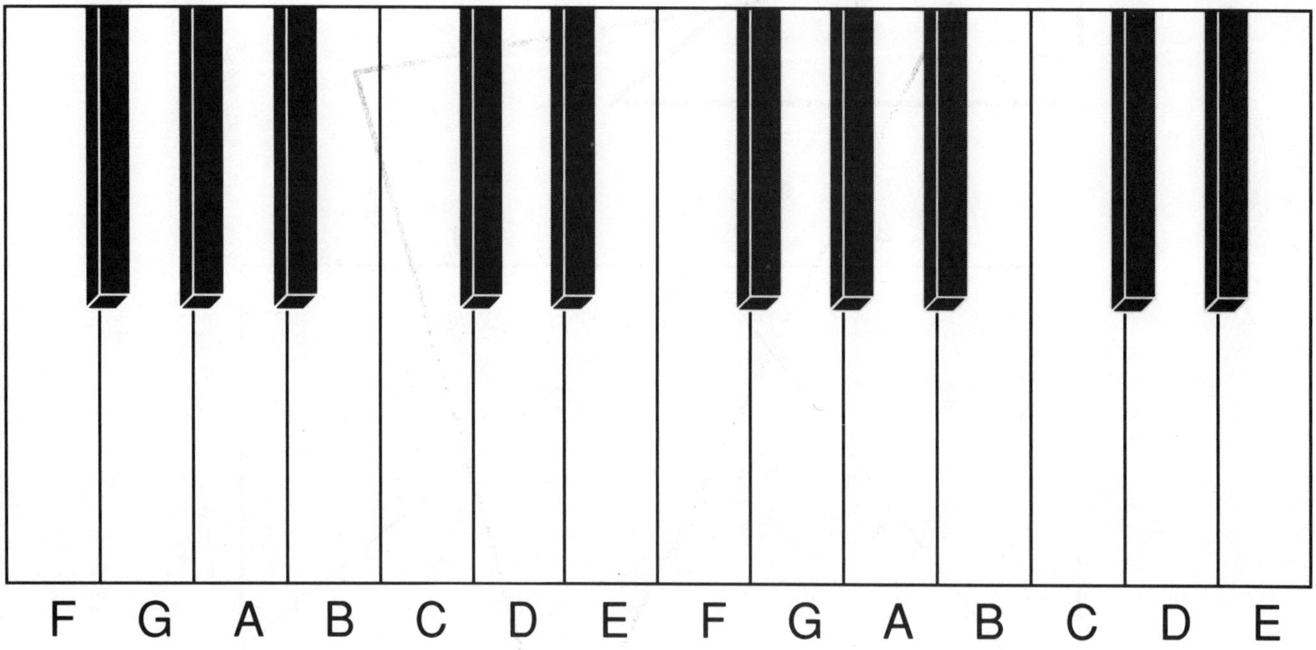

Question 22

By drawing an altitude and measuring lengths, find the area of this triangle to the nearest inch.

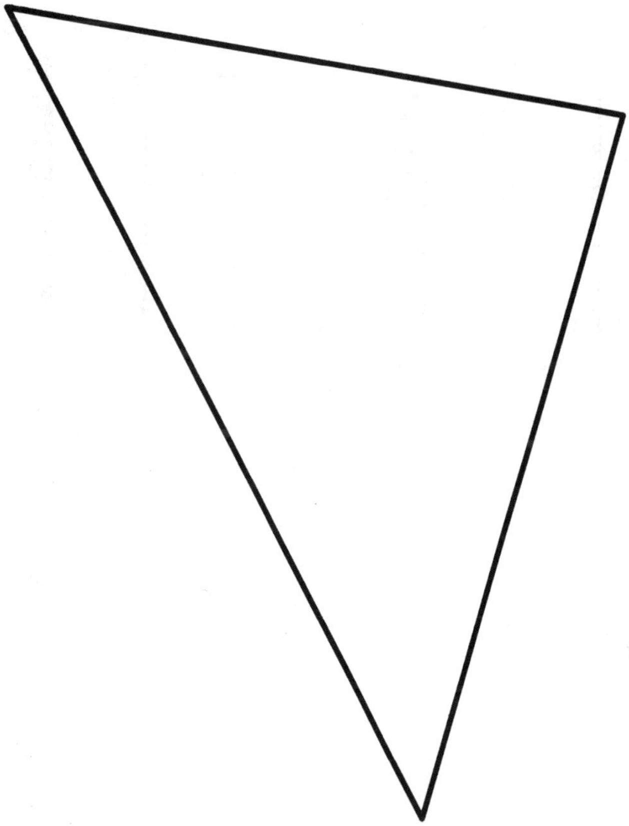

The Family of Real Numbers

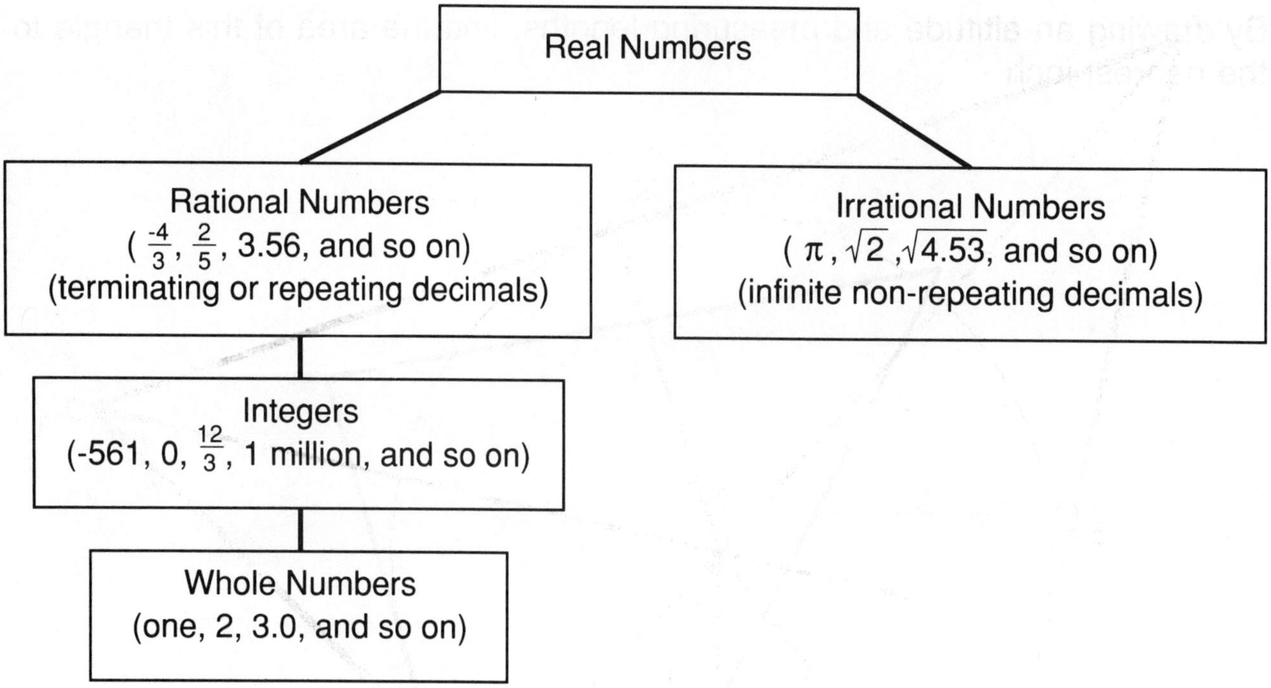

Question 29

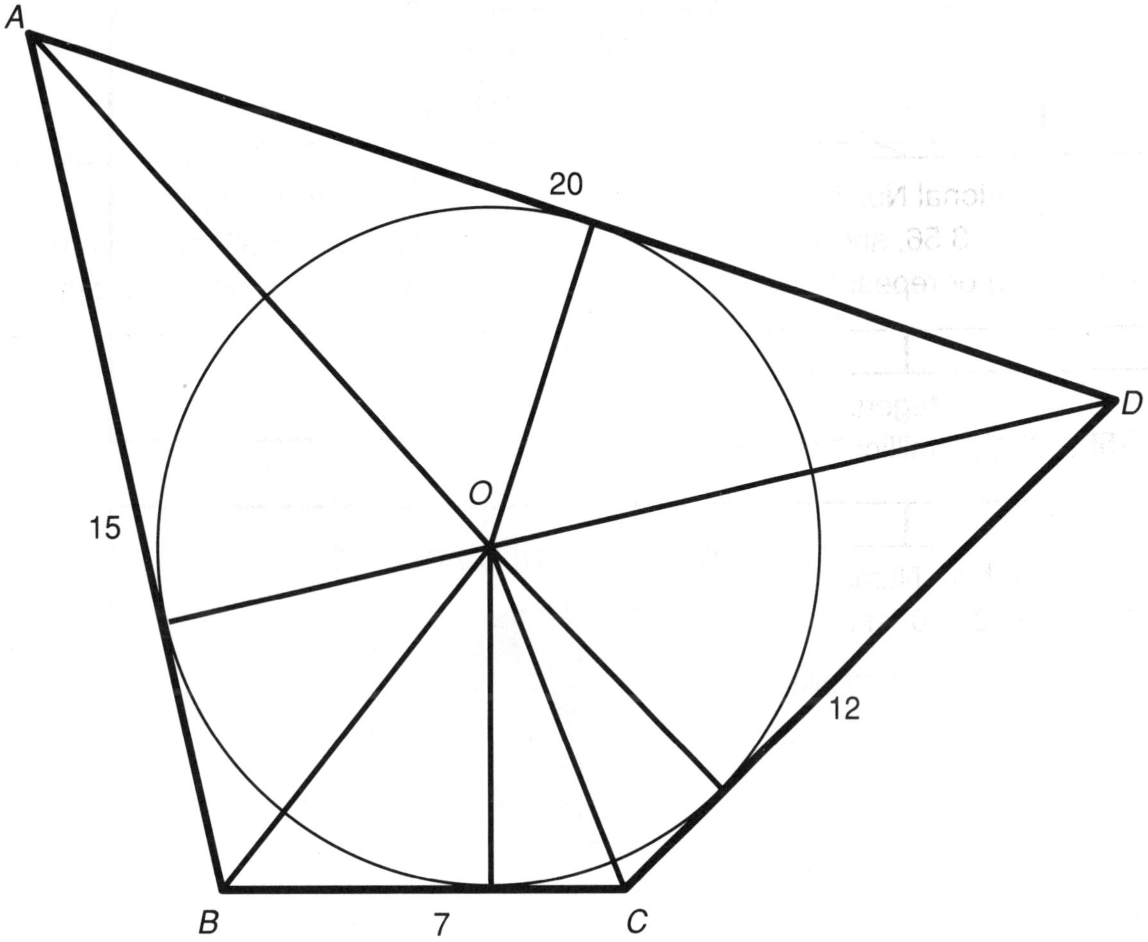

Circle Graphs

Choice	Number of Students Picking that Choice
Ice cream	40
Cake	20
Fruit	15
None	5

Dessert choices of 80 people

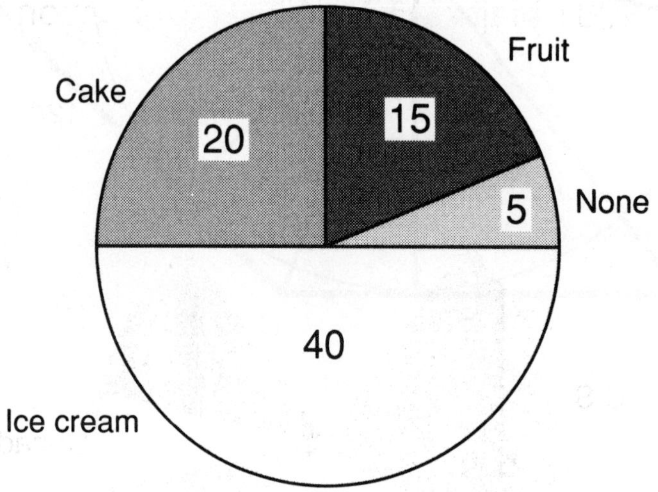

Example

1981 Leading wheat producers	
U.S.S.R.	90
U.S.	75
China	60
India	35
Canada	20
France	20
Total	300

Country	w/300	w/300•360°
U.S.S.R.	.30	108°
U.S.	.25	90°
China	.20	72°
India	.12	42°
Canada	.07	24°
France	.07	24°

1981 Major World Wheat Production

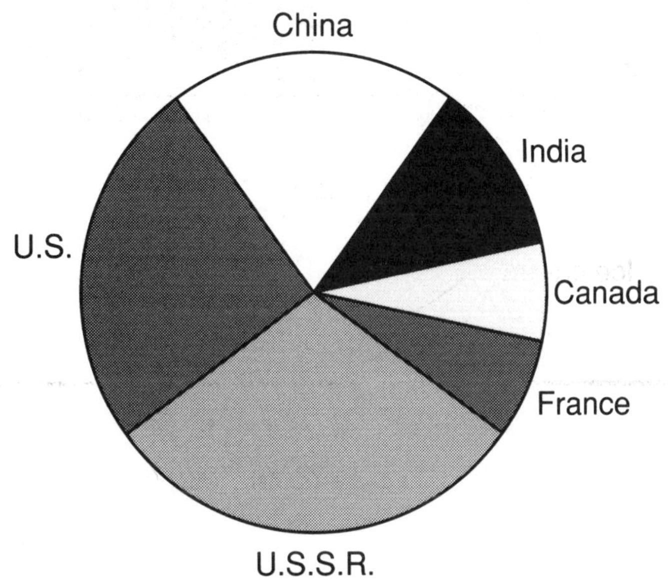

Question 13

From the circle graph:

a. Which age bracket has the fewest people?
b. Which age bracket has the most people?
c. Which age is missing?
d. How many people are between the ages of 6 and 19?

The USA has 2 million more senior citizens than teenagers, according to this month's statistics that put the population at 236.4 million. Latest population projections by age.

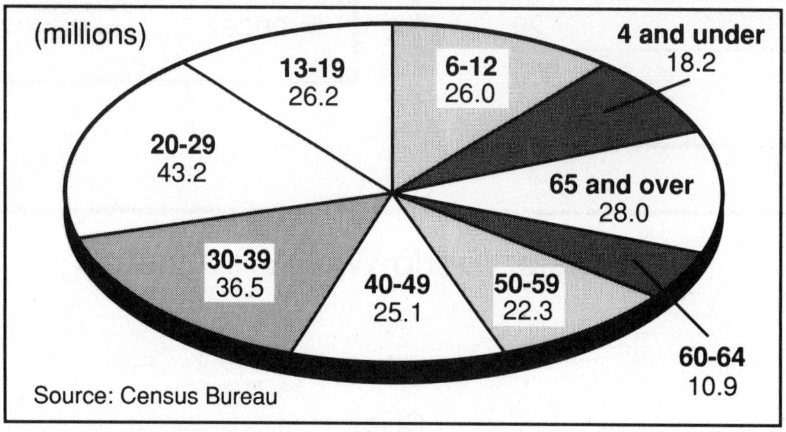

What other questions can be answered from the circle graph?